Operational Effects of Geometric and Access Management 2018

Transportation Research Record: Journal of the Transportation Research Board Volume 2672, Issue 17

ISBN: 978-1-5108-8181-5

Printed from e-media with permission by:

Curran Associates, Inc.
57 Morehouse Lane
Red Hook, NY 12571

Some format issues inherent in the e-media version may also appear in this print version.

Published by Sage Publications USA

ISBN (Print) 978-1-5108-8181-5
ISBN 2018 Printed Set (All Issues) 978-1-5108-7735-1

Printed by Curran Associates, Inc. (2019)

Additional copies of this publication are available from:

Curran Associates, Inc.
57 Morehouse Lane
Red Hook, NY 12571 USA
Phone: 845-758-0400
Fax: 845-758-2633
Email: curran@proceedings.com
Web: www.proceedings.com

Volume 2672 Number 17 December 2018

TRR
JOURNAL OF THE TRANSPORTATION RESEARCH BOARD

Contents

Operational Effects of Geometric and Access Management 2018

Articles

Transportation Officials, Washington, D.C.
K. Jane Williams, *Acting Administrator, Federal Transit Administration, U.S. Department of Transportation*
Paul Zukunft *(Admiral, U.S. Coast Guard), Commandant, U.S. Coast Guard, Washington, D.C.*

---

* Membership as of February 2019.

TRB

TRANSPORTATION RESEARCH BOARD

CORE SPONSORS

Transportation Departments of the 50 States and the District of Columbia

Federal Government

U.S. Department of Transportation
- Federal Aviation Administration (FAA)
- Federal Highway Administration
- Federal Motor Carrier Safety Administration (FMCSA)
- Federal Railroad Administration (FRA)
- Federal Transit Administration
- National Highway Traffic Safety Administration (NHTSA)
- Office of the Assistant Secretary of Transportation for Research and Technology (OST-R)

U.S. Air Force Civil Engineer Center (AFCEC)
U.S. Coast Guard (USCG)
U.S. Department of Energy
U.S. Department of the Interior (DOI)
U.S. Environmental Protection Agency (EPA)

Nongovernmental Organizations

American Public Transportation Association (APTA)
Association of American Railroads (AAR)
California Air Resources Board

The National Academies of
SCIENCES · ENGINEERING · MEDICINE

The **National Academy of Sciences** was established in 1863 by an Act of Congress, signed by President Lincoln, as a private, nongovernmental institution to advise the nation on issues related to science and technology. Members are elected by their peers for outstanding contributions to research. Dr. Marcia McNutt is president.

The **National Academy of Engineering** was established in 1964 under the charter of the National Academy of Sciences to bring the practices of engineering to advising the nation. Members are elected by their peers for extraordinary contributions to engineering. Dr. C. D. Mote, Jr., is president.

The **National Academy of Medicine** (formerly the Institute of Medicine) was established in 1970 under the charter of the National Academy of Sciences to advise the nation on medical and health issues. Members are elected by their peers for distinguished contributions to medicine and health. Dr. Victor J. Dzau is president.

The three Academies work together as the National Academies of Sciences, Engineering, and Medicine to provide independent, objective analysis and advice to the nation and conduct other activities to solve complex problems and inform public policy decisions. The Academies also encourage education and research, recognize outstanding contributions to knowledge, and increase public understanding in matters of science, engineering, and medicine.

Learn more about the National Academies of Sciences, Engineering, and Medicine at **www.national-academies.org.**

The **Transportation Research Board** is one of seven major programs of the National Academies of Sciences, Engineering, and Medicine. The mission of the Transportation Research Board is to increase the benefits that transportation contributes to society by providing leadership in transportation innovation and progress through research and information exchange, conducted within a setting that is objective, interdisciplinary, and multimodal. The Board's varied activities annually engage about 7,000 engineers, scientists, and other transportation researchers and practitioners from the public and private sectors and academia, all of whom contribute their expertise in the public interest. The program is supported by state transportation departments, federal agencies including the component administrations of the U.S. Department of Transportation, and other organizations and individuals interested in the development of transportation.

Learn more about the Transportation Research Board at **www.TRB.org**.

TRR
JOURNAL OF THE TRANSPORTATION RESEARCH BOARD

Article

Transportation Research Record
2018, Vol. 2672(17) 1–10

DOI: 10.1177/0361198118795668
journals.sagepub.com/home/trr
SAGE

Access Separation at Interchanges: Examining Crash Rates on the Cross-Street and in the Transition Areas from Full to Partial Access Control

Timothy P. Barrette[1], Jacob Warner[2], Patricia Thompson[3], and Peter T. Savolainen[4]

Abstract

Access control remains an important concern to roadway agencies as the spacing of at-grade access points significantly affects both the safety and operational performance of highways. Significant variability exists with respect to policies dictating where the first access points may occur in the vicinity of interchanges. This study examines two important spacing criteria, which include: (1) the distance from the freeway off-ramp to the first access on the interchanging arterial roadway; and (2) the distance from a freeway on-ramp to the first at-grade intersection in areas with partial access control. To investigate these relationships, a series of crash prediction models were estimated that examined how crash risk changed with respect to access spacing while controlling for the effects of traffic volume and other pertinent characteristics. The results illustrate that at cross-streets near ramp terminal intersections, the crash rate generally increases as the distance to the nearest access point decreases. The same trend was also true of freeway transition areas. Ultimately, this research illustrates a complex relationship that exists among the proximity of the exit point of the controlled access facility and the adjacent access point, the volume of traffic along the roadway, and the volume of traffic at the access point.

Access management refers to a series of techniques used to control vehicular access between roadways and adjacent land parcels. This access is provided through signalized and unsignalized intersections, driveways, and median openings. In general, the introduction of any such access point results in the deterioration of both the operational and safety performance of a road facility (*1*). The magnitude of these impacts varies based upon factors such as the traffic volumes of both the roadway and the access connection, as well as geometric design factors such as available sight distance and turning radii. Transportation agencies have some control over these factors and one of the most important techniques for effectively managing access involves the establishment of minimum criteria for the distance between consecutive access points.

Access management is particularly important in the vicinity of interchanges, where the number, location, and types of access points can have substantive impacts on traffic operations and safety. Poor access management can lead to several undesirable impacts, including reduced speeds, increased travel time and delay, and a heightened risk of traffic crashes, injuries, and fatalities (*1*). Significant variability exists with respect to policies dictating where the first access points may occur in the vicinity of interchanges. This study examines two important spacing criteria that include:

1. The distance from the freeway off-ramp to the first access on the interchanging arterial roadway; and
2. The distance from a freeway on-ramp to the first at-grade intersection in areas with partial access control.

The first concern is related to the location of access points on the cross-street corridor near the ramp terminal as shown in Figure 1. These areas are generally served by ramp connections, which terminate at signalized,

[1]Texas A&M Transportation Institute, College Station, TX
[2]Department of Civil, Construction, and Environmental Engineering, Iowa State University, Ames, IA
[3]AECOM, Southfield, MI
[4]Department of Civil and Environmental Engineering, Michigan State University, East Lansing, MI

Corresponding Author:
Address correspondence to Peter T. Savolainen: pete@msu.edu

Figure 1. Cross-street corridor near ramp terminal intersection.

Table 1. Summary of Minimum Access Spacing Standards or Guidelines (Gluck et al. [*4*])

State/Province	Rural	Urban
1. Alabama	300 ft to access	100 ft to access
2. Alberta	425 m from signal to access	Same
3. California	150 m from ramp to access	Same
4. Illinois	125 m minimum distance from ramp to nearest intersection	Same
5. Iowa	200 m rural primary highway, 100 m other road or street	50 m urban
6. Kentucky	300 ft to access	100 ft to access
7. Maryland	Based on geometrics, speeds, volumes, presence of signals, and queuing	Same
8. North Dakota	AASHTO guidelines	AASHTO guidelines (100 ft)
9. Ohio	600 ft for diamond interchange, 1,000 ft for cloverleaf	
10. Oregon	200 ft from frontage road, 500 ft from ramp (suggested)	Same
11. Pennsylvania	AASHTO guidelines (300 ft)	AASHTO guidelines (100 ft)
12. South Carolina	500 ft desirable, 300 ft minimum	300 ft desirable, 150 ft minimum
13. Texas	AASHTO guidelines (300 ft)	AASHTO guidelines (100 ft)
14. Utah	300 ft to access	150 ft to access
15. Virginia	200 ft from entrance ramp	Same
16. West Virginia	300 ft to access	100 ft to access
17. Washington	300 ft to access	300 ft to access
18. Wisconsin	1,000 ft to access, (500 ft minor roads)	500 ft to access
19. Wyoming	300 ft to access	150 ft to access

stop-controlled, yield-controlled, or free-flowing intersections. The location of the first access point from the end of the exit ramp is dependent on various factors including the type of traffic control, as well as the traffic volume, speed limit, level of development, and type of access.

Early design guidance allowed for the first access point to occur as close as 100 ft from the ramp terminal (*2*). Table 1 provides a summary of minimum access spacing standards and guidelines from NCHRP Report 420 (*3*). These guidelines, which were drawn from various agencies in the United States and Canada, show spacing minimums to vary from 100 to 700 ft in urban areas and from 300 to 1,000 ft in rural areas. Recently, access policies have generally changed and many states have since adopted more stringent guidelines for access.

Increased spacing for access points that are located within close proximity to ramp terminals has become important as the extant literature clearly documents issues that arise on these corridors. These include complicated weaving movements and complex signal operations, which can lead to congestion and increased crash risks (*3*, *4*). These issues may be exacerbated when drivers are transitioning from an uninterrupted flow facility (i.e., freeway) to a lower class cross-street. Depending on the exact interchange configuration, drivers may have to complete several maneuvers after exiting a freeway, such as merging, changing lanes, and navigating through an access point (*5*).

A second access management concern addressed as a part of this study is the location of the first at-grade

Figure 2. Analysis corridor at expressway transition area.

intersection or driveway as a fully access-controlled freeway transitions to a partially controlled expressway. Figure 2 illustrates one such example where an at-grade intersection occurs east of the I-35/US-30 interchange near Ames, Iowa. Although some guidance exists as to spacing between consecutive freeway ramps, guidance for these expressway transition areas is more limited.

The purpose of this study is two-fold. First, the research aims to examine how crash risk varies on cross-street corridors based upon the distance at which the first access point is located relative to the ramp terminals. Currently, the Iowa Highway Primary Access Management Policy (*6*) establishes minimum distances where access rights shall be acquired along the intersecting public road or street at interchanges. In rural and fringe areas, the first access should be located a minimum of 600 ft from the point of ramp bifurcation. In built-up areas, a minimum distance of 150 ft is required, with 300 ft desired. The Iowa Department of Transportation (Iowa DOT) maintains the right to acquire more or less than these specified access rights after considering the severity of damage to adjacent properties, traffic volumes, and other safety factors. When the interchange is constructed as a half-diamond or partial cloverleaf, access may be permitted directly opposite the ramp connection to the primary road. The second objective of this study is to examine how crash risk varies on expressways where transitions occur between full and partial access control. In these cases, access is generally limited to one entrance for contiguous highway frontage not exceeding 1,000 ft. However, explicit guidance is lacking regarding the transition from full to partial access control.

Literature Review

Various research studies have investigated issues related to the spacing of the first access point relative to the ramp intersection. A recurring theme in the extant literature is the effect of access point proximity on congestion. One research project aimed at using computer simulations to develop guidelines for access point location in Florida found that signals should be placed at minimum ¼-mi from the ramp bifurcation point, whereas ½ mi should be used in cases where high development is expected (*7*). A recent study utilized a mixed-integer non-linear model to develop non-traditional lane assignment through pavement markings and signal timings in an urban setting to minimize congestion at the off-ramp intersection (*8*).

In addition to previous efforts that have focused on the operational effects of access point location, the extant literature contains examples of studies focusing on the safety implications of access point proximity to ramp bifurcation points as well. A Virginia study utilized data from 186 access road sections to estimate negative binomial and least square linear regression models to examine the effect of access point spacing relative to the ramp bifurcation point (*9*). The findings of this study were very pronounced: an increase in the spacing from 0 to 300 m equated to an eight-fold decrease in the crash rate, whereas an increase from 90 to 180 m resulted in a 50% decrease in the crash rate. A subsequent report documented the similarity of the results between the negative binomial and linear regression approaches (*10*).

Efforts have been made by researchers to combine the safety and operational effects of access point proximity near interchanges into monetary value. A study in Florida examined the relationship between distance to the first access point on a frontage road and the effect on congestion and crashes on the adjacent freeway (*11*). Key findings of this study included marked improvements in traffic flow when the distance to the first access point on a frontage road increased from 200 to 600 ft, a general decreasing trend in crash frequency as the distance to the access point increased, and large benefit–cost ratios associated with the purchasing of land near interchanges ahead of development occurring.

In the *Highway Safety Manual* (HSM), a series of crash modification factors are presented for interchanges, however they focus on the conversion of at-grade intersections to grade separations (*12*). Safety performance functions, or crash prediction models, are presented for a variety of facility types to help road agencies assess the safety at a variety of roadway facilities, however freeways are not included. In preparation for the forthcoming second edition of the HSM, NCHRP Project 17-45 developed crash frequency prediction models for nearly all components of a freeway, including ramp terminal intersections (*13*); however, the cross-street area beyond the ramp terminal intersection and the transition area from full to partial access control were not explicitly addressed.

Data Collection

As a part of this study, extensive datasets were created for: (1) the cross-street corridor at every service (i.e., freeway-to-arterial) interchange throughout the state of Iowa; and (2) the transition areas where a freeway on-ramp is followed by an at-grade public intersection or private driveway. Details of these respective datasets follow.

Ramp Terminal Cross-Street Database

Interchanges were initially identified using an attribute query of the Iowa DOT Geographic Information Management System (GIMS) database to identify any segment classified as a ramp. This initial review identified a total of 406 interchanges, which included service interchanges, where a freeway connects with a cross-street of lower functional class, as well as system interchanges connecting two freeways. A subsequent quality control review was conducted, which involved the collection of pertinent information such as the interchange configuration (e.g., diamond, cloverleaf), the type of traffic control on the interchange ramps (e.g., signalized, stop-controlled), and whether the interchanging road was another freeway or an arterial roadway. The attribute query utilized to identify interchanges also identified non-interchange ramps, such as slip ramps at four-leg intersections. As the scope of this study was focusing on interchanges from uninterrupted flow facilities to cross-streets, these sites were also excluded from the analysis.

This process led to the identification of 380 service interchanges. These interchanges included 724 arterial corridors, which were the subject of an extensive manual review. As a part of this review, the following information was collected for each corridor:

- Distance to first access point (ft)
- Side of road where first access point was located (near, far)
- Classification of first access point (signalized intersection, unsignalized intersection, driveway, field access)
- Allowable movements at first access point (full-movement, left-in, left-out, right-in, right-out)
- Median type at point of ramp bifurcation and at first access point (undivided, divided, two-way left-turn lane)
- Median width at point of ramp bifurcation point (ft)
- Median width at first driveway (ft)
- Police-reported crash data from the Iowa DOT crash database
- Roadway traffic volume and geometric information from the Iowa DOT GIMS
- Volumes estimated based on driveway classification.

Volumes were estimated for the driveway access points using the Institute of Transportation Engineers (ITE) *Trip Generation Manual* (*14*). For each classification determined by the researchers, the number of trips were calculated using the average values listed in the manual. Due to the difference in weekday and weekend traffic, the annual average daily traffic (AADT) for the classification was calculated using the average of the sum of the Saturday trips, Sunday trips, and five times the weekday trips.

As these volumes are representative of the average example of the category, in some cases the AADT estimated for the driveway exceeded that of the road the driveway was located on. In these instances, if the driveway was not opposite another roadway, the volume on the driveway was constrained to be equal to the volume on the main road.

Descriptive statistics for the cross-street dataset are shown in Table 2. The table contains information about the roadway up to the first access point, as well as more detailed information about that first access point, including estimated volumes, classification, allowable movements, and distance from the point of ramp bifurcation. Approximately half of the access points were comprised of commercial driveways, residential driveways, or farm access roads. Roughly 12% of the first access points were signalized, with the remainder being stop-controlled locations.

This dataset was used to examine the relationship between crashes and distance to the first access point independent of anything that occurred after that access point. The 7 years of data (2010 to 2016) collected for the cross-street dataset represent 724 corridor segments. Collectively, these segments experienced an average of 1.23 crashes per year up to a maximum of 34 crashes. After compiling the dataset, a preliminary investigation

Table 2. Descriptive Statistics for Ramp Terminal Cross-Street Data

Variable	Mean	Standard deviation	Minimum	Maximum
First access point distance (ft)	750.99	641.70	1.00	4,329.10
Cross-street annual average daily traffic	6,010.23	6,953.54	10.00	44,900.00
Off-ramp annual average daily traffic	1,803.77	2,363.93	10.00	16,700.00
Access point annual average daily traffic	1,095.14	2,349.75	0.00	20,300.00
Number of lanes	2.85	1.17	2.00	9.00
First access = driveway	0.49	0.50	0.00	1.00
First access = stop-controlled intersection	0.39	0.49	0.00	1.00
First access = signalized intersection	0.12	0.32	0.00	1.00
Signalized ramp	0.22	0.41	0.00	1.00
Yield-controlled ramp	0.04	0.19	0.00	1.00
Total crashes	1.23	2.78	0.00	34.00

Note: n = 724 segments.

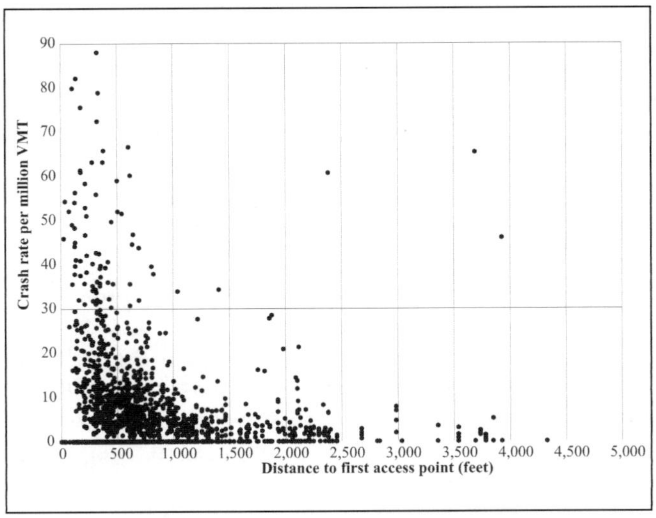

Figure 3. Crash rate per million VMT versus distance to first access point on cross-street.

was conducted to examine the relationship between crash rate (per million VMT) and the distance to the first access point on the cross-street traveling away from the point of ramp bifurcation. Figure 3 displays a plot of crash rate versus this distance to the first access. This plot shows a strong decreasing trend and clearly illustrates the heightened crash risk that occurs, particularly within the first few hundred feet of the ramp.

Expressway Transition Area Dataset

A similar dataset was created for the expressway transition areas. In recent years, the state of Iowa has been expanding its expressway network. The GIMS does not have a field that specifically identifies expressways. Consequently, an extensive manual review was conducted that utilized a map of existing and planned expressways obtained from Iowa DOT. All expressways throughout the state were identified that were operational during the entirety of the study period from 2010 through 2016. Upon identifying these highways, satellite imagery was reviewed to identify areas where these roadways transition from full to partial access control. The transition area dataset again utilized 7 years' crash data (2010 to 2016), based on data availability at the onset of the study. Ultimately, this process yielded a dataset consisting of 57 corridors over the 7-year period, which are described in Table 3.

Figure 4 illustrates the crash rate of the freeway transition areas relative to the length of the transition area. In contrast to the ramp terminal sites, there are no extremely short transition areas, however, the data still exhibit a decreasing trend in crash rate as segment length increases.

Statistical Analysis

To discern the impacts of access spacing on the rate of motor-vehicle crashes, a series of crash prediction models were estimated to determine how crash risk changes based on the distance to the first access point while considering the effects of other pertinent predictors. Separate models were estimated for: (1) the distance to the first access point from the point of ramp bifurcation at a service ramp terminal; and (2) the distance to the first at-grade access point where a full-access control section transitions to partial access control.

These regression equations take the form of generalized linear models. As crash data are comprised of non-negative integers, traditional regression techniques (e.g., ordinary least-squares) are generally not appropriate. Given the nature of such data, the Poisson distribution has been shown to provide a better fit and has been used widely to model crash frequency data. In the Poisson model, the probability of road segment i experiencing y_i crashes during a one-year period is given by Equation 1,

Table 3. Descriptive Statistics for Expressway Transition Area Data

Variable	Mean	Standard deviation	Minimum	Maximum
Distance to first access point (ft)	3,589.10	1,715.54	938.00	8,539.80
Annual average daily traffic	9,258.94	4,604.28	2,322.40	23,800.00
Intersection with median gap	0.79	0.41	0.00	1.00
No access points in first mile	0.18	0.38	0.00	1.00
3 leg access point	0.27	0.45	0.00	1.00
4 leg access point	0.56	0.50	0.00	1.00
First access point is public road	0.59	0.49	0.00	1.00
Signalized access point	0.04	0.18	0.00	1.00
Roundabout access point	0.02	0.13	0.00	1.00
First access point within ¼ mi	0.04	0.18	0.00	1.00
First access point between ¼ and ½ mi	0.35	0.48	0.00	1.00
First access point between ½ and ¾ mi	0.30	0.46	0.00	1.00
First access point between ¾ and 1 mi	0.14	0.35	0.00	1.00
First access point greater than 1 mi	0.18	0.38	0.00	1.00
Total (non-animal) crashes	0.80	1.17	0.00	8.00

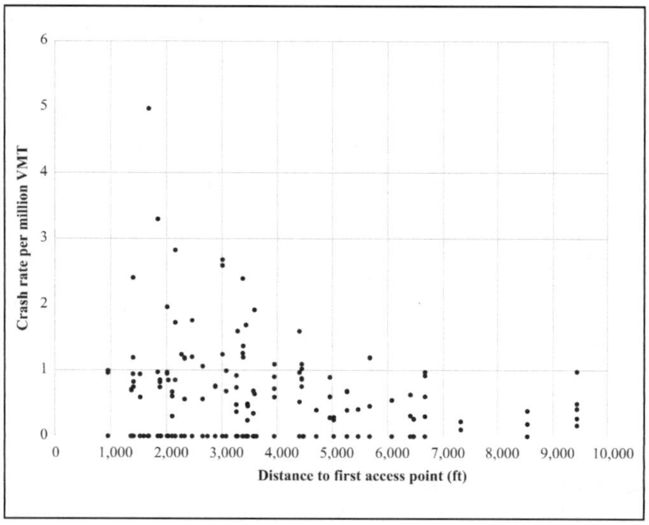

Figure 4. Crash rate per million VMT vs. distance to first access point in expressway transition area.

$$P(y_i) = \frac{EXP(-\lambda_i)\lambda_i^{y_i}}{y_i!} \quad (1)$$

where $P(y_i)$ is probability of segment i experiencing y_i crashes and λ_i is the Poisson parameter for segment i, which is equal to its expected number of crashes per year, $E[y_i]$. Poisson models are estimated by specifying the Poisson parameter λ_i (the expected number of crashes per period) as a function of explanatory variables, the most common functional form being given by Equation 2,

$$\lambda_i = exp(\beta X_i) \quad (2)$$

where X_i is a vector of explanatory variables (e.g., AADT, distance to first access) and β is a vector of estimable parameters.

One potential limitation of this model is the underlying assumption of the Poisson distribution that the variance is equal to the mean. As such, the model cannot handle overdispersion wherein the variance is greater than the mean. Overdispersion is common in crash data and may be caused by data clustering, unaccounted temporal correlation, model misspecification, or ultimately by the nature of the crash data, which are the product of Bernoulli trials with unequal probability of events (*15*). Overdispersion is generally accommodated through the use of negative binomial models (also referred to as Poisson-gamma models). The negative binomial model is derived by rewriting the Poisson parameter for each segment as shown in Equation 3,

$$\lambda_i = EXP(\beta X_i + \varepsilon_i) \quad (3)$$

where $EXP(\varepsilon_i)$ is a gamma-distributed error term with mean 1 and variance α. The addition of this term allows the variance to differ from the mean as shown in Equation 4:

$$VAR[y_i] = E[y_i] + \alpha E[y_i]^2 \quad (4)$$

The negative binomial model is preferred over the Poisson model as the latter cannot handle overdispersion and, as such, may lead to biased parameter estimates (*16*). If the overdispersion parameter (α) is equal to zero, the negative binomial reduces to the Poisson model. Estimation of λ_i can be conducted through standard maximum likelihood procedures.

Since repeated observations were conducted on each segment over time (i.e., five years of data are included), there is temporal correlation among these crash counts. Failure to account for this correlation may result in biased or inefficient parameter estimates due to site-specific factors that are not accounted for in the model.

Table 4. Results of Negative Binomial Model for Crashes at Ramp Terminal Cross-Street

Parameter	Estimate	Standard error	T-stat	P-value
Intercept	−2.427	0.459	−5.286	<0.001
Var(Intercept)	0.551			
Log(annual average daily traffic of crossroad)	0.576	0.057	10.090	<0.001
Log(annual average daily traffic of off-ramp)	0.352	0.053	6.623	<0.001
Log(annual average daily traffic of first access point)	0.072	0.022	3.266	0.001
Log(distance to first access point)	−0.686	0.033	−20.908	<0.001
Signalized interchange (1 if yes; 0 otherwise)	0.363	0.120	3.032	0.002
Signalized first access (1 if yes; 0 otherwise)	0.551	0.127	4.336	<0.001
First access is stop-controlled intersection (1 if yes; 0 otherwise)	0.312	0.090	3.465	0.001
Number of lanes	0.111	0.042	2.625	0.009
Overdispersion parameter	0.542	0.142	3.817	<0.001

To address this concern, random-effect models were estimated wherein the constant term in Equation 5 is allowed to vary across segments as follows:

$$\beta_{0i} = \beta_0 + \omega_i \qquad (5)$$

where ω_i is a randomly distributed random effect for segment i and all other variables are as previously defined. When estimating crash prediction models for road segments, the natural log of the segment length is generally treated as an offset variable, which means its parameter estimate is constrained to be equal to 1. This introduces an assumption that crashes will increase proportionately with segment length.

Results and Discussion

In order to develop useful guidance as to where the first access point should be located for each facility type, separate crash prediction models were estimated. As a part of this analysis, various model formulations were considered. This included examining the change in crash rate based on 100-ft intervals, as well as treating distance as a continuous variable assuming a constant rate of change in crashes. Ultimately, this investigation showed consistent trends that were best estimated by treating distance as a continuous predictor variable.

As data was collected on an annual basis for these analyses, a random effects model was estimated to account for temporal correlation within the same sites (i.e., certain locations may tend to experience more or fewer crashes than other similar locations due to variables that are not included in the dataset). First, the results of the crash prediction model for the cross-street data is shown in Table 4.

The results show that the random effects model provided significantly improved fit as compared with a naïve-pooled negative binomial model (p-value < 0.001). This result suggests that although various parameters were found to significantly affect the crash rate at these sites, there are still important unobserved site-specific factors that also affect the crash rate.

The model estimated for this study examined the volume of the cross-street, as well as the volumes of the freeway exit ramp, and the first access point. As expected, each of these volumes was found to have a substantial impact on crash frequency along the corridor. In examining the results from Table 4, a 1% increase in volume on the crossroad results in a 0.59% increase in crashes, whereas 1% increase in volume on the off-ramp and access point results in an increase of 0.34% and 0.07%, respectively. It is interesting to note that the parameter associated with the first access point traffic volume is relatively small. This suggests that the effect of access point volume increases rapidly at low volumes and then plateaus as access point volume increases.

The primary variable of interest in this study was the distance to the first access point from the point of ramp bifurcation. The estimated parameter associated with this characteristic was negative, which indicates that by increasing the distance to the first access point, the expected crash rate between the ramp bifurcation point and first access point decreases. Figure 5 presents the results of this analysis in graphical format. This provides two important illustrations: (1) the degree to which crash risk decreases the further an access point is located from the ramp; and (2) the impact of cross-street volume on crash risk. Ultimately, these results provide guidance as to appropriate distance thresholds for access point location. Figure 5 shows crash rates to decrease significantly up to a distance of approximately 600 ft, after which more gradual declines occur prior to a plateau at around 1,300 ft.

A similar analysis was conducted for the expressway transition areas. This analysis considered all of the freeway transition corridors in the dataset, even those that did not have an access point within 1 mi of the last freeway interchange. The latter segments, in essence, provided baseline estimate of the crash rate

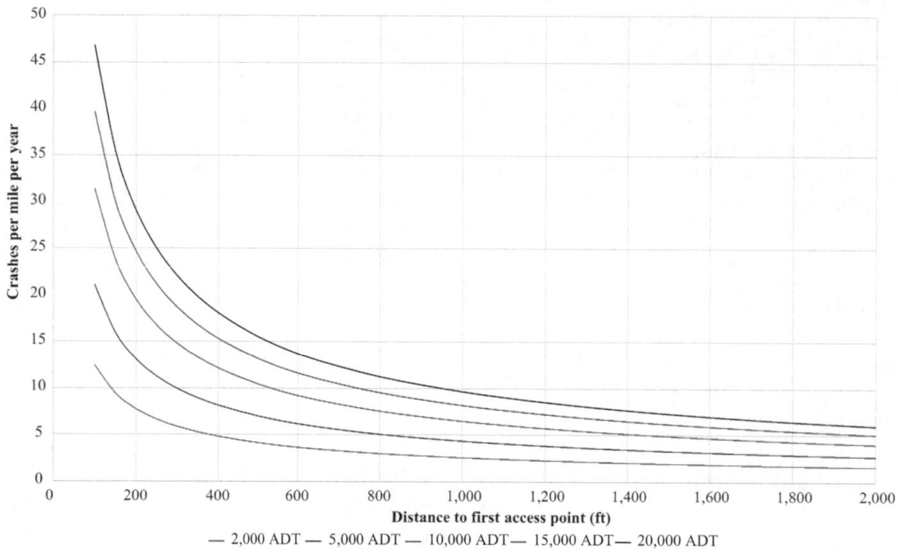

Figure 5. Crashes per mile per year on cross-street versus distance to first access.

Table 5. Results of Negative Binomial Model for Crashes in Expressway Transition Area

Parameter	Estimate	Standard error	*T*-stat	*P*-value
Intercept	−5.601	1.419	−3.947	<0.001
Variance(Intercept)	1.149			
Log(annual average daily traffic)	0.616	0.156	3.940	<0.001
Log(distance to first access point)	−0.335	0.141	−2.371	0.018
Signalized first access (1 if yes; 0 otherwise)	0.864	0.281	3.081	0.002
Overdispersion parameter	0.298	0.143	2.084	0.038

for a similar facility with no access points. The goal of this analysis was to determine any correlation between crash rate and the distance to the first access point, as well as other factors that could impact the crash rate. These factors include whether the first access point was a public road versus a private driveway, as well as whether a median gap was present. Results of this model are shown in Table 5 and presented in graphical format in Figure 6.

Collectively, these results show there is significant correlation between the crash rate at the first access point in a freeway transition area and the distance between the access point and the freeway interchange. These curves follow a similar trend to the cross-street dataset; however, the overall number of crashes is significantly lower given the higher design standards and relatively low traffic volumes of much of the expressway network. The crash rate tapers rapidly until a distance of between 1,000 and 2,000 ft from the last exit ramp, with a plateau occurring at around 3,000 ft.

Similar to the results of the access point spacing at on-ramp terminal cross-streets, these results also show that higher traffic volumes contribute to higher crash rates. The results also show that the presence of a signalized intersection is related to a higher crash rate. This can be attributed to the fact that motorists driving on a freeway would not expect to see a signal without having seen cross traffic previously on that roadway.

Conclusions

Ultimately, the results of this study provide important guidance to inform decisions as to where access points should be located in consideration of impacts on crash risk. Visual inspection of the data showed a consistent underlying relationship between crash rate, traffic volume, and the distance to the first access point at both the crossroad corridors and the expressway transition areas. An inverse relationship was found between crashes per mile and distance to first access on each facility type.

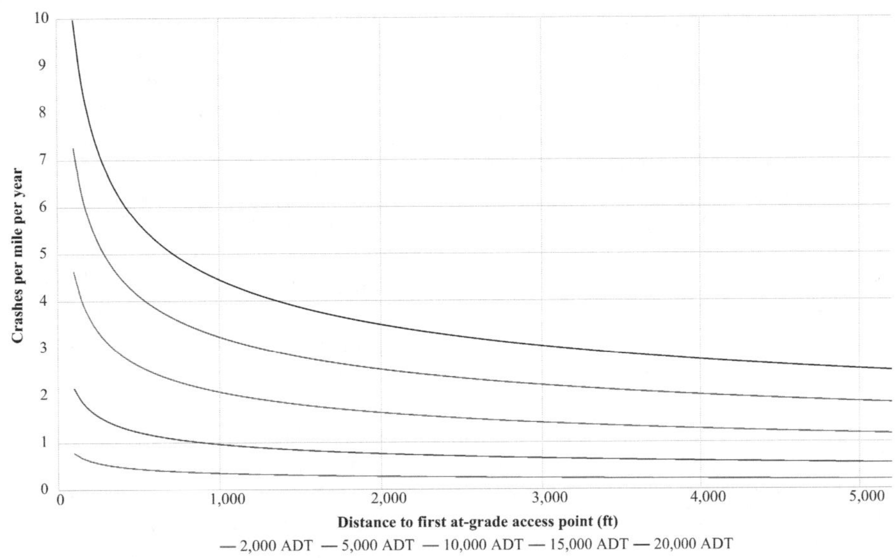

Figure 6. Crashes per mile per year in transition area versus distance to first access.

Intuitively, an increasing relationship between traffic volume and crash rate was also observed.

Traffic volume on the cross-street was shown to have a larger effect on crash rate than the volumes of the off-ramp or access point. However, each of these was shown to have a significant impact on safety. One valuable contribution of this study to the extant literature is the incorporation of access point volumes into the functional form of the crash prediction model. Several previous studies have used loosely defined, arbitrary terms such as major intersection and minor intersection in lieu of better volume metrics (*9*, *10*). The study dataset included measured, estimated or both measured and estimated AADT data for all public roadways, as well as estimates of traffic volumes for the driveways based on ITE Trip Generation (*14*). Consequently, the framework utilized in this study provides a more flexible, robust basis for decision making as it relates to the location and spacing of access points.

Though safety will virtually always be adversely affected by the introduction of access points along a corridor, these effects can be mitigated by locating access points as far from the preceding access as possible. Crash rates on cross-street corridors are pronounced when the distance between the ramp terminal and access point is less than 1,000 ft, with dramatic increases when access points are located within 600 ft and, particularly, 300 ft.

This study was somewhat limited in terms of geometric characteristics present on the cross-street, due in part to the rural nature of many interchanges in the state of Iowa. Future work could aim to incorporate data from other, more densely populated areas, which may provide better insight as to specific geometric characteristics effecting safety near ramp terminal intersections. Additional information could also be collected for comparable corridors that are not ramp-adjacent. This would allow for an analysis to provide insight as to the safety of these corridors relative to the broader population of roadways in Iowa, which could potentially result in the determination of a base level of safety to which the ramp-adjacent corridors could be compared.

Author Contributions

The authors confirm contribution to the paper as follows: study conception and design: PS; data collection: TB, JW, PT; analysis and interpretation of results: PS, TB, JW, PT; draft manuscript preparation: PS, TB, JW. All authors reviewed the results and approved the final version of the manuscript.

References

1. Williams, K. M., V. G. Stover, K. Dixon, and P. Demosthenes. *Access Management Manual.* Transportation Research Board, Washington, D.C., 2014, pp. 529.
2. American Association of State Highway and Transportation Officials. *A Policy on Geometric Design of Highways and Streets.* Washington, D.C.: AASHTO, 1991.
3. Gluck, J., and M. R. Lorenz. *NCHRP Report 420: Impacts of Access Management Techniques.* Transportation Research Board of the National Academies, Washington, D.C., 2010.
4. Gluck, J., H. S. Levinson, and V. Stover. *NCHRP Synthesis 404: State of the Practice in Highway Access*

Management. Transportation Research Board of the National Academies, Washington, D.C., 1999.
5. Florida Department of Transportation. *Median Handbook*. Florida Department of Transportation, Tallahassee, Fla., 2014.
6. Iowa Department of Transportation. *Iowa Primary Highway Access Management Policy*. Iowa Department of Transportation, Ames, IA., 2012.
7. Washburn, S. S., and A. Kondyli. *Development of Guidelines for Driveway Location and Median Configuration in the Vicinity of Interchanges*. Report No. TRC-FDOT-036-2006. Florida Department of Transportation, Tallahassee, Fla., 2006.
8. Zhao, J., and Y. Liu. Integrated Signal Optimization and Non-Traditional Lane Assignment for Urban Freeway Off-Ramp Congestion Mitigation. *Transportation Research Part C*, Vol. 73, 2016, pp. 219–238.
9. Rakha, H., A. M. Flintsch, M. Arafeh, G. Abdel-Salam, D. Dua, and M. Abbas. *Access Control Design on Highway Interchanges*. Report No. VTRC 08-CR7. Virginia Department of Transportation, Richmond, Va., 2008.
10. Rakha, H., M. Arafeh, G. Abdel-Salam, F. Guo, and A. M. Flintsch. *Linear Regression Crash Prediction Models: Issues and Proposed Solutions*. Report No. VT-2008-02. Virginia Department of Transportation, Richmond, Va., U.S. Department of Transportation, Washington D.C., 2010.
11. Williams, K. M., H. Zhou, L. Hagen, and W. Farah. *Benefit-and-Cost Analysis of Strategic Acquisition of Limited Access Right-of-Way near Interchanges*. 6th National Access Management Conference, Kansas City, Mo., 2004.
12. American Association of State Highway and Transportation Officials. *Highway Safety Manual*, 1st ed. Washington, D.C.: AASHTO, 2010.
13. Bonneson, J. A., S. Geedipally, M. P. Pratt, and D. Lord. *NCHRP Project 17-45: Safety Prediction Methodology and Analysis Tool for Freeways and Interchanges*. Transportation Board of the National Academies, Washington, D.C., 2012.
14. Institute of Transportation Engineers (ITE). *Trip Generation Manual*, 8th ed. Washington, D.C., 2008.
15. Lord, D. Modeling Motor Vehicle Crashes Using Poisson-gamma Models: Examining the Effects of Low Sample Mean Values and Small Sample Size on the Estimation of the Fixed Dispersion Parameter. *Accident Analysis and Prevention*, Vol. 38, No. 4, 2006, pp. 751–766.
16. Lord, D., and F. Mannering. The Statistical Analysis of Crash-Frequency: A Review and Assessment of Methodologies. *Transportation Research, Part A: Policy and Practice*, Vol. 44, No. 5, 2010, pp. 291–305.

The Standing Committee on Access Management (AHB70) peer-reviewed this paper (18-00009).

TRR
JOURNAL OF THE TRANSPORTATION RESEARCH BOARD

Article

Transportation Research Record
2018, Vol. 2672(17) 11–21

DOI: 10.1177/0361198118795997
journals.sagepub.com/home/trr
SAGE

An Examination of the Safety of Signalized Intersections in Consideration of Nearby Access Points

Megat-Usamah Megat-Johari[1], Bahareh Bazargani[2], Trevor J. Kirsch[3], Timothy P. Barrette[4], and Peter T. Savolainen[1]

Abstract

Effective access management is critical to the safe and efficient operation of roadway corridors. Access management is particularly important near signalized intersections, where adjacent driveways exacerbate the crash risks created by conflict points at the primary intersection. This study examines the relationship between crash risk and access spacing in the vicinity of 415 signalized urban intersections. The frequency of access points near each signalized intersection of interest was collected to examine the differences in experienced crash frequency over a five-year period. Additional information about each access point was collected, including details of turning movement restrictions and estimated driveway volumes. A series of buffers of varying radii were considered to examine how crash rates varied based upon the spatial proximity of access points to each signalized intersection. In addition, traffic volume and roadway geometric information was collected and utilized as a part of this evaluation. A series of crash prediction models were estimated to discern how crash rates varied based upon the access point density on the intersection approaches along the major (i.e., higher volume) roadway. As the number of access points increased, the crash rates also increased. This relationship was consistent in direction, but varied in magnitude, depending on the size of the buffers, which varied in radius from 50 ft to 600 ft at 50-ft intervals. The number of crashes by location was also shown to vary based upon the number of travel lanes, the number of intersection legs, and the land use type of the adjacent driveways.

Access management is paramount in facilitating safe and efficient roadway operations. Effective access management policies allow state and local governments to control effectively the manner in which vehicles gain access to and from the public roadway network from private facilities. These policies establish guidelines as to the location and spacing of intersections, driveways, and median openings. Access management strategies can have positive effects on traffic safety and operations by reducing the number of crashes and increasing roadway capacity, thereby reducing delay. These objectives can be achieved by applying various strategies, such as increasing the distance between access points, limiting the number of access points from properties adjacent to the traveled way, and limiting traffic conflicts through the use of medians, diverters, and other roadway design elements. In contrast, poor access management can impede traffic flow, create additional congestion, and increase the potential for traffic conflicts, thereby increasing the risk of traffic crashes.

Most of the crashes that occur near access points are usually attributed to conflicts between multiple motorized vehicles. These conflicts often occur due to a vehicle attempting to turn across a conflicting traffic stream at an intersection or driveway, or from vehicles attempting to exit the major road at another access point. Vehicles turning left or right into access points are typically exposed to right-angle and rear-end crashes. These conflicts occur along the midblock portion of road segments, but are more pronounced near intersections, particularly when other adjacent access points are present. The close spacing between intersections and driveways does not provide enough clearance for vehicles to turn into and out of driveways without potentially conflicting with vehicles attempting to maneuver through the intersection.

[1]Department of Civil and Environmental Engineering, Michigan State University, East Lansing, MI
[2]Department of Civil, Construction and Environmental Engineering, Iowa State University, Ames, IA
[3]WSP USA, Detroit, MI
[4]Texas A&M Transportation Institute, Texas A&M University, College Station, TX

Corresponding Author:
Address correspondence to Peter T. Savolainen: pete@msu.edu

Previous research has illustrated that as the number of midblock access points per mile increase, the crash rate is expected to increase as well (*1*). While the impact of midblock access point density has been examined in great detail, the spacing of access points in close vicinity to intersections needs further examination (*1*, *2*).

This study examines how crash rates at signalized intersections vary based upon the number and type of adjacent access points. The analysis focuses on a sample of signalized intersections located on the Iowa primary (i.e., state-maintained) road network. A total of 415 signalized urban intersections were selected from several Iowan cities and were included in this analysis. Intersections from urbanized areas were utilized specifically due to the prevalence of access points near these intersections. The frequency of access points near the signalized intersections of interest was calculated to examine the effect of this variable on crash frequency. Additional information about the access points was collected, which included the type of development associated with access, which allowed for the estimation of site-generated traffic volume as well as the presence of turning movement restrictions for each driveway. Additionally, crash data, traffic volume, and roadway geometric information were collected and utilized in this study. A series of buffers (concentric circles from the center of the intersection) were created to examine how the relationship between crashes and access point density varies spatially depending upon the radius of interest. These buffers range from 50 ft to 600 ft in radius in 50-ft increments. Using all the listed information, crash prediction models were estimated to determine the effect of access point density on crashes at intersection approaches.

Literature Review

Access density is a widely-accepted technique that has been extensively examined in studies related to traffic safety, operational impacts, and planning (*3*). Previous research has shown that crash rates increase when the total number of access points or the density of access points increases (*4*). Another study found that density of driveways and unsignalized minor streets, and different median types are significantly correlated with crash frequency (*5*). However, most of these studies evaluated the safety effects of access points at a segment level, with only a few of these studies evaluating the effects of access point density at an intersection level.

One such study examined the effect of corner clearance (i.e., the distance from the intersection to the first access point) based on a sample of 300 signalized intersections in an urban area (*6*). The results of this research determined that the average length of corner clearance had a negative correlation with the number of crashes. More specifically, the further the first driveway was spaced from the intersection, the greater the resultant traffic safety (i.e., less traffic crashes). Moreover, this study also found that different types of land developments had different effects on the corner clearance. In general, residential areas had longer corner clearance intervals when compared with commercial areas. Consequently, the results showed that residential areas were expected to experience a lower number of crashes than their commercial area counterparts.

An additional study found that the number of commercial driveways within 250 ft of a signalized intersection was associated with an increase in the expected number of crashes (*7*). The results estimated an increase of approximately 7% in the expected number of crashes with every additional commercial driveway within 250 ft from the intersection. A similar study determined that corner clearance had negative correlation with crash frequency (*8*). The results showed that for every 100 ft of additional corner clearance, three crashes could be avoided. Further research examined 144 signalized intersections in the state of Utah based on the safety performance of signalized intersections when considering the number of commercial driveways within the intersection functional area (*9*). From this study, it was found that the number of commercial driveways within this area had a positive correlation with the total number of crashes and resultant crash rate.

In general, studies concerning access management have demonstrated a positive effect on both highway safety and operational characteristics, including a roadway's ability to maintain traffic flow as well as travel speed, all without affecting the business outcome of the adjacent land (*10*). Based on prior research, well-managed arterial roadways have been shown to reduce traffic crashes by 40% to 50% in comparison with poorly managed arterials on a per vehicle mile basis. Additionally, arterial highways have been shown to have significantly higher travel speeds as well as better traffic levels-of-service on well-managed corridors (*10*). Lastly, when considering the resultant effects on roadside businesses, studies have demonstrated that well-managed roadways have little to no impact on the financial outcome of the businesses as well as the overall customer satisfaction at the establishment (*11–13*).

In the state of Iowa, between 5% and 10% of all crashes occur at commercial driveways (*10*). Most of these occur along arterials within municipalities. In recent years, nearly 25% of these crashes have occurred in the Des Moines metropolitan area. Due to the Des Moines metropolitan area having many miles of older arterials that were constructed before access management was considered, the area is a prime candidate for improved access management.

Some work has been done to assess the safety impacts of access point density through the development of a safety performance function (SPF). SPFs are statistical models used to predict crash frequency at a given site based on site-specific exposure measures: traffic volume and length of the segment. In this model, other variables are also included to obtain more reliable results than those available when relying solely on the exposure measures to predict crash frequency (*2*, *11*). These SPFs are developed through statistical regression models which use historical crash data (*14*). Much of the extant literature focuses on negative binomial regression as the framework utilized to estimate models for predicting crash frequency due to the typical overdispersion of crash data (*2*, *6*, *15*).

In order to quantify the safety performance of driveway locations on arterial highways, the Oregon Department of Transportation developed a variety of arterial-based SPFs. Based on an analysis of the crash reporting fidelity for driveway-related crashes, the SPFs were developed for non-intersection locations. The crashes considered in the developed functions were arterial-related on both urban and rural state highways in Oregon. In order to validate the model, the spatial transferability of the SPF was tested, as well as the spatial–temporal transferability, the individual coefficient stability, and the overall significance. To measure the performance of the developed equations, the SPFs were applied to an urban arterial and tested using the previously mentioned metrics. Ultimately, the developed SPFs performed well under the spatial transferability condition, resulting in statistically significant values; however, the functions did not perform well under any other testing condition (*16*).

This study specifically examines the impact of driveway and unsignalized intersection density and proximity relative to signalized intersections through the estimation of SPFs. The data utilized in this research comes from urban locations across the state of Iowa.

Data Collection

The principal focus of this study is to examine the safety implications of driveway density near urban intersections on the primary road network. To accomplish this task, data was integrated from the following sources:

- Iowa Department of Transportation (Iowa DOT) Geographic Information Management System (GIMS)
- Iowa DOT Intersection Database
- Iowa DOT Crash Database
- Google Earth satellite imagery

Due to the georeferenced nature of the various data sources, the data were ultimately integrated into a coherent, spatially accurate format using ArcGIS.

Iowa DOT Geographic Information Management System (GIMS)

Three unique datasets from the GIMS portal were utilized in this study: Traffic, Road Info, and Direct Lane. Each file contains a certain type of information pertaining to a specific roadway segment. The Traffic dataset contains relevant traffic parameters, including a measurement of annual average daily traffic (AADT). The Road Info dataset outlines important roadway parameters, including surface width and lane count. The Direct Lane dataset includes various characteristics related to the countermeasures and infrastructure of the roadway. Roadway segments on the primary network were also identified using these files. Ultimately, the three datasets were used to identify the major and minor roads at each intersection based on traffic volumes.

Iowa DOT Intersection Database

The intersection information utilized in this study was obtained from a database developed by the Iowa DOT. Due to possible differences in performance, this study focuses specifically on non-ramp, at-grade intersections, which were identified from the larger population of all Iowa intersections. The information contained in the intersection database includes a variety of categorical information, including number of legs, general shape (T- and Y-intersections, roundabouts), approach direction (two-way or one-way), type of median (if present), and type of intersection control. In this study, each intersection is grouped based on the number of legs present and the type of intersection control as follows: four-leg intersections with signalized control and three-leg intersections with signalized control. Figure 1 shows an aerial view of a typical four-leg intersection as well as an identification of the first and second access points on the major road.

Iowa DOT Crash Database

Crash data were obtained from the Iowa DOT crash database for a period of five years from 2011 to 2015. For each intersection, the number of crashes that occurred in a given year were identified based on a range of buffer distances. These buffers ranged from 50 ft to 600 ft in radius at 50-ft intervals. Two subsets of crashes were identified within these respective radii along the major (i.e., higher volume) roadway: total crashes and driveway-related crashes. The driveway-related crashes were identified based upon the fields available on the Iowa police crash report form.

Manually Collected Data

In this study, manual data collection was necessary in order to identify all access points along the primary road

Figure 1. Example access points within 300 ft of a signalized intersection.

for the selected intersections. This was done using heads-up digitizing in ArcGIS, where the locations of driveways, field accesses, alleys, median openings, and intersections were recorded. In addition, the type of land use associated with each driveway was determined, which allowed for the estimation of site-generated traffic volumes using the Institute of Transportation Engineers (ITE) Trip Generation Manual. In addition to land use type, the area surrounding each intersection was classified as residential or commercial. The presence of any turning movement restrictions (i.e., prohibited left-turns in, prohibited left-turns out, or right-in/right-out only) for each access point was also determined. The number of access points on each corridor was also used to calculate the access point density within each buffer range.

Data Summary

To obtain a suitable crash prediction model that examines the relationship between crash frequencies and all variables of interest, a total sample of 415 signalized intersections is utilized in this study. The descriptive statistics of predictor variables are presented in Table 1. These data summarize the frequency of driveways, average traffic volumes of all driveways, and the percentage of driveways that prohibit left-turn movements that are located within radii of 50 to 600 ft from the center of each signalized intersection. Traffic volumes ranged from 853 to more than 37,000 vehicles per day along the primary roadways that were the focus of this study, with a mean of 17,083. The intersecting minor roadways had an average volume of 4,982 vehicles per day. Other geometric characteristics of interest included the speed limits, the numbers and types of lanes, and the presence of medians on both the major and minor road.

One of the primary factors of interest is the number of crashes that occurred over the five-year analysis period. Separate summaries are provided for total crashes, as well as for driveway-related crashes based upon information provided from the police-reported crash data as shown in Table 2. One complementary objective of this study was to discern how crash rates varied based upon the subsets of crashes that were considered. Prior research has suggested some issues related to the accuracy of police-reported crash data, particularly when attempting to identify those crashes that may be intersection- or driveway-related.

Statistical Methods

After all the data had been integrated appropriately, multiple statistical analyses were conducted to determine the effect of access point density on crashes near the primary network intersections. Additional variables related to roadway geometry and site characteristics are also included. Traffic crashes are random, discrete, and non-negative integers. There are two types of commonly used count models in the development of SPFs: Poisson and negative binomial regression models. The Poisson regression model is usually considered first when modeling

Table 1. Descriptive Statistics for Site Characteristics

Variable	Min.	Max.	Mean	Std dev.
Number of driveways within 600 ft	0	29	4.72	5.45
Number of driveways within 550 ft	0	26	4.23	4.97
Number of driveways within 500 ft	0	23	3.76	4.43
Number of driveways within 450 ft	0	21	3.33	3.98
Number of driveways within 400 ft	0	20	3.00	3.63
Number of driveways within 350 ft	0	16	2.66	3.25
Number of driveways within 300 ft	0	15	2.26	2.83
Number of driveways within 250 ft	0	13	1.80	2.33
Number of driveways within 200 ft	0	10	1.21	1.68
Number of driveways within 150 ft	0	7	0.70	1.11
Number of driveways within 100 ft	0	4	0.29	0.60
Number of driveways within 50 ft	0	2	0.10	0.32
Average AADT of Driveways within 600 ft	0	9100	765	975
Average AADT of Driveways within 550 ft	0	9100	731	955
Average AADT of Driveways within 500 ft	0	9100	717	971
Average AADT of Driveways within 450 ft	0	9100	719	1028
Average AADT of Driveways within 400 ft	0	9100	714	1085
Average AADT of Driveways within 350 ft	0	9100	673	1048
Average AADT of Driveways within 300 ft	0	9100	649	1089
Average AADT of Driveways within 250 ft	0	9100	605	1093
Average AADT of Driveways within 200 ft	0	9100	576	1132
Average AADT of Driveways within 150 ft	0	9100	499	1156
Average AADT of Driveways within 100 ft	0	9100	329	1090
Average AADT of Driveways within 50 ft	0	15486	246	1223
Percentage of prohibited left-turn within 600 ft	0	1	0.60	0.42
Percentage of prohibited left-turn within 550 ft	0	1	0.62	0.42
Percentage of prohibited left-turn within 500 ft	0	1	0.63	0.43
Percentage of prohibited left-turn within 450 ft	0	1	0.65	0.43
Percentage of prohibited left-turn within 400 ft	0	1	0.65	0.43
Percentage of prohibited left-turn within 350 ft	0	1	0.67	0.43
Percentage of prohibited left-turn within 300 ft	0	1	0.68	0.43
Percentage of prohibited left-turn within 250 ft	0	1	0.69	0.43
Percentage of prohibited left-turn within 200 ft	0	1	0.71	0.43
Percentage of prohibited left-turn within 150 ft	0	1	0.76	0.42
Percentage of prohibited left-turn within 100 ft	0	1	0.84	0.37
Percentage of prohibited left-turn within 50 ft	0	1	0.91	0.28
AADT for major road	853	37300	17083	6804
AADT for minor road	13	23800	4982	4192
Speed limit (mph) for major road	25	55	37.04	7.14
Speed limit (mph) for minor road	20	60	31.18	8.59
Number of lanes on major road	2	9	4.61	1.16
Number of lanes on minor road	2	7	2.91	1.16
Presence of median on major road (1 if yes; 0 otherwise)	0	1	0.57	0.50
Presence of median on minor road (1 if yes; 0 otherwise)	0	1	0.16	0.37
Land development (1 if commercial; 0 if residential)	0	1	0.81	0.39
Presence of left-turn lane on major road (1 if yes; 0 otherwise)	0	1	0.57	0.50
Presence of left-turn lane on minor road (1 if yes; 0 otherwise)	0	1	0.16	0.37
Number of legs (1 if four leg; 0 if three leg)	0	1	0.80	0.40

crash data. The probability of the number of crashes, y_i occurring at a given intersection i during a specific period is

$$P(y_i) = \frac{e^{-\lambda} \cdot \lambda_i^{y_i}}{y_i!} \quad (1)$$

where λ_i is the Poisson parameter for intersection i, which for this study is the expected number of crashes that could occur within the radius of the buffers at a given time period for intersection i. The expected number of crashes can be expressed as

$$\lambda_i = EXP(\beta_1 X_1 + \beta_2 X_2 + \ldots + \beta_n X_n) \quad (2)$$

where X_1 through X_n represent site-specific explanatory variables (such as traffic information and roadway geometry), and β_1 through β_n represent parameter estimates

Table 2. Descriptive Statistics for Crashes by Type and Buffer Radius

Variable	Min.	Max.	Mean	Std dev.
Total crashes, 600-ft radius	0	152	35.59	29.14
Total crashes, 550-ft radius	0	151	33.66	27.91
Total crashes, 500-ft radius	0	135	32.23	26.98
Total crashes, 450-ft radius	0	129	30.49	25.91
Total crashes, 400-ft radius	0	124	27.59	24.05
Total crashes, 350-ft radius	0	113	24.90	22.30
Total crashes, 300-ft radius	0	108	23.14	21.14
Total crashes, 250-ft radius	0	103	21.58	20.09
Total crashes, 200-ft radius	0	101	20.12	19.21
Total crashes, 150-ft radius	0	96	18.19	17.97
Total crashes, 100-ft radius	0	93	16.61	17.13
Total crashes, 50-ft radius	0	88	14.60	15.91
Target crashes (driveway-related), 600-ft radius	0	39	1.61	3.80
Target crashes (driveway-related), 550-ft radius	0	39	1.49	3.69
Target crashes (driveway-related), 500-ft radius	0	39	1.35	3.41
Target crashes (driveway-related), 450-ft radius	0	38	1.24	3.13
Target crashes (driveway-related), 400-ft radius	0	37	1.07	2.87
Target crashes (driveway-related), 350-ft radius	0	37	0.97	2.71
Target crashes (driveway-related), 300-ft radius	0	36	0.85	2.46
Target crashes (driveway-related), 250-ft radius	0	35	0.70	2.14
Target crashes (driveway-related), 200-ft radius	0	27	0.55	1.71
Target crashes (driveway-related), 150-ft radius	0	20	0.38	1.27
Target crashes (driveway-related), 100-ft radius	0	12	0.26	0.85
Target crashes (driveway-related), 50-ft radius	0	4	0.16	0.52

from the regression model. One of the assumptions when using the Poisson model is that the mean number of crashes is equal to the variance. It is typical of crash data that the variance is greater than the mean, which is known as overdispersion. To overcome the overdispersion typical of crash data, a negative binomial model is frequently used. The negative binomial model is a generalized version of the Poisson model, as demonstrated in Equation 3

$$\lambda_i = EXP(\beta_1 X_1 + \beta_2 X_2 + \ldots + \beta_n X_n + \varepsilon_i) \quad (3)$$

where $EXP(\varepsilon_i)$ is gamma-distributed with a mean equal to one and a variance α (overdispersion parameter). With the inclusion of the overdispersion parameter, the variance when using a negative binomial regression model can be expressed as

$$Var(y_i) = E(y_i) + \alpha E(y_i)^2 \quad (4)$$

The marginal effect of each variable on the identified crashes were calculated using Equation 5. This equation gives a change in percentage in the number of crashes when there is a change in the value of the independent variable. The equation can be expressed as

$$\Delta\lambda = 100\left(e^{\beta_n X_n} - 1\right) \quad (5)$$

where $\Delta\lambda$ is the percentage change in the number of crashes.

Results and Discussion

In this study, two sets of crashes (total and driveway-related) on the major road intersection approaches were analyzed to determine the effects of access point density on crash frequency. Table 2 shows that the number of driveway crashes that most of the intersections experienced was zero (even when the 600-ft radius buffer was considered), as evidenced by the mean value. The first 100-ft buffer for driveway-related crashes resulted in all intersections collectively experiencing less than 100 crashes per year, which is the minimum number of crashes that the sample should have to develop a reliable SPF (*17*).

As a part of this study, analyses were conducted for buffers in 50-ft increments from 50 to 600 ft The subsequent discussion focuses primarily on the analyses using the largest (i.e., 600-ft) buffer. Table 3 presents the results of the negative binomial model for driveway target crashes within 600 ft of the intersection. From these results, it is evident that the number of driveways is positively correlated with crash frequency. The number of crashes would be expected to increase by 9.3% when one driveway is added within the 600 ft distance of the intersection. Based on previous studies, this result shows a similar trend where crashes will increase when driveway density increases. Other predictor variables also demonstrate the same trend from previous studies. Increases in traffic volume on the major and minor roads of 1% were

Table 3. Negative Binomial Model Results for Driveway Target Crashes with 600-ft Buffer

Variable	Estimate	Std error	Z-value	P-value
Intercept	−11.767	2.008	−5.861	<0.001
Natural logarithm of major road AADT	0.810	0.205	3.958	<0.001
Natural logarithm of minor road AADT	0.250	0.085	2.932	0.003
Natural logarithm of average AADT of driveways	0.133	0.036	3.702	<0.001
Number of driveways	0.089	0.016	5.623	<0.001
Percentage of prohibited left turn	−0.647	0.228	−2.830	0.005
Residential land development (base)	–	–	–	–
Commercial land development	1.285	0.283	4.535	<0.001
Overdispersion parameter	1.248	0.185	6.731	<0.001

Table 4. Negative Binomial Model Results for Total Crashes with 600-ft Buffer

Variable	Estimate	Std error	Z-value	P-value
Intercept	−4.030	0.800	−5.035	<0.001
Natural logarithm of major road AADT	0.486	0.088	5.535	<0.001
Natural logarithm of minor road AADT	0.222	0.036	6.086	<0.001
Number of driveways	0.040	0.006	6.180	<0.001
Residential land development (base)	–	–	–	–
Commercial land development	0.374	0.091	4.090	<0.001
Major road—2 lanes (base)	–	–	–	–
Major road—3 lanes	0.715	0.210	3.405	0.001
Major road—4 lanes or more	0.522	0.173	3.023	0.003
Overdispersion parameter	0.456	0.033	13.621	<0.001

associated with 0.81% and 0.25% increases in driveway-related crashes, respectively. Similarly, a 1% increase in the estimated driveway volume was associated with a 0.13% increase in crashes. Access management strategies, such as turning movement restrictions, were considered, with left-turn prohibition showing a significant impact on crash frequency. On average, crashes were reduced by 48% if left-turn movements were fully prohibited within the buffer area. Note that this variable is reflective of the percentage of driveways where turning movements were prohibited, as most intersections restricted turning movements at a subset of driveways. As turning movements are restricted, the number of conflict points is reduced, reducing the probability of involvement in a driveway-related crash. When considering the type of development around the intersections, the number of crashes was 261% greater when the intersection was located in a commercial area when compared with a residential area.

From a practical standpoint, it is difficult to discern whether crashes are ultimately due to the presence of a driveway or not in many circumstances. Focusing on total crashes mitigates potential concerns regarding the accuracy of the crash report codes that designate whether the crash was intersection- or driveway-related. As driveway-related crashes are a relatively small subset of total crashes, and there is some difficulty in identifying crashes that are truly driveway-related, the effect of driveway presence on total crashes is also of particular interest from a research and policy standpoint. The remainder of the discussion focuses on the results from the analysis of total crashes.

Based on the estimates in Table 4, six variables were associated with higher crash frequencies. As expected, traffic volume and crashes have a positive correlation. Increasing the traffic volume on the major or minor road would increase the number of predicted crashes. A 1% increase in traffic volume on the major road would result in a 0.48% increase in the number of crashes. For the minor road, a 1% increase in traffic volume would result in a 0.22% increase in the number of crashes.

The results showed that the number of driveways within 600 ft of the intersection influenced the total number of crashes. For each additional driveway within this buffer distance, the number of crashes would increase by 4.10%. Increasing the number of driveways increases the number of conflict points between the primary roadway and the entering/exiting driveway traffic.

Intersections in commercial areas experienced 45.4% more crashes on average than those in residential areas. This finding suggests that the driveway volume in commercial areas was greater than the residential areas, resulting in more frequent traffic conflicts. Compared with commercial areas, it is likely that drivers in residential areas have a familiarity with the roads surrounding their neighborhood, thus, they are familiar with the geometric characteristics of the road and may be less likely to be involved in a crash.

The number of lanes also had an impact on the number of crashes, with an increase in number of lanes resulting in an increase in crashes. Roads with three lanes experienced approximately twice as many crashes when compared with roads with two lanes. Roads with more than three lanes were subject to 69% more crashes, on average, when compared with roads with two lanes. Initially, this result may seem counterintuitive, as one might expect a consistent increase in the number of crashes as the number of lanes increases. One possibility for this observation was that locations with three lanes (in which a two-way left-turn lane becomes a dedicated turn lane immediately adjacent to the intersection) were less likely to utilize split phase signal timings. Additionally, the two-way left-turn lane itself naturally introduces conflicts between vehicles travelling in opposing directions, and is often used as a refuge for vehicles exiting driveways, which likely generates greater crash frequencies. Turning movement channelization features (which provide space for vehicles to decelerate adjacent to the through moving traffic) were likely more prevalent on the roadways with more than three lanes, thus reducing the likelihood of a rear-end collision.

Overall, crashes increased as the number of lanes increased. Drivers need to change lanes more frequently and plan their movements further in advance with denser access point spacing. Therefore, an increase in the number of lanes increases the likelihood of motorists engaging in traffic conflicts related to diverging, merging, and weaving between vehicles, which increases the risk of a crash occurring (*15*). The number of legs at the intersection was not found to be a significant determinant of crash frequency. One possible reason for this finding is the small sample size of three-leg intersections in comparison with four-leg intersections as shown in Table 1.

Lastly, the posted speed limits for both the major and minor roadways did not show a significant relationship with the safety performance of the intersection. However, it should be noted the average speed limits for the major and minor roads were 37.0 mph and 31.2 mph, respectively. Additional investigation is warranted, particularly at higher posted limits, to understand better the nature of the speed-safety relationship. This issue is complicated by the fact that access density and speed limit are generally directly or indirectly related to one another based on functional classification and various design controls.

Table 5 presents a summary of the results of the models that were estimated for each of the twelve buffer radii for total crashes. By utilizing buffers of various sizes, the effect of spatial proximity of the access points relative to the signalized intersections can be considered. Results from these tables outline a demonstrable increase in the magnitude of the parameter estimates associated with driveway count. As the buffer size decreases, the rate of crashes tends to increase more rapidly. These results suggest that the closer a driveway is to an intersection, the greater the crash risk for motorists.

This increase in crash risk is intuitive, as less space is available for drivers to change lanes and navigate through the intersection as desired, which may result in risky operator behavior which increases the probability that a rear-end or sideswipe collision will occur. Note that the Highway Safety Manual (HSM) states that it is undesirable to have any access points located within a 250 ft radius of an intersection. This finding is similar to the estimated models presented in these two tables, which demonstrate that the number of crashes will increase as the first driveway moves closer to the intersection.

Figure 2 graphically displays the estimated number of crashes for each buffer based on the different driveway counts, while holding all other independent variables constant. Each buffer is representative of the different SPFs documented in Table 5. Note that the safety performance of each buffer demonstrates an increase in crash frequency as driveway count increases. In other words, adding driveways to the approaches of the intersections of interest increased the crash frequency of driveway and intersection target crashes, regardless of the size of the buffer. Interestingly, as the size of buffer increases, the rate of increase in crashes per additional access points decreases (though the total frequency of crashes still increases with each access). This demonstrates that the further the access point is located from the intersection, the less the expected impact on crash risk.

Conclusion

Based on the results of this analysis, access point density has been shown to have a significant effect on various crashes types in the immediate vicinity of signalized intersections. Crashes were shown to consistently increase with the number of driveways located along the primary road near signalized intersections.

The effect of access point proximity to intersections was examined through the estimation of a series of negative binomial models. Separate analyses were conducted for buffer areas between 50 ft and 600 ft in 50-ft intervals. From these results, it was observed that limiting access point density and restricting the proximity in which an access point can be located relative to a signalized intersection is critical to safe intersection operation. These results provide support for the development of policies or guidelines by state and local governments to

Table 5. Negative Binomial Model Results for Total Crashes for Buffers from 50 to 600 ft

	50-ft radius			100-ft radius			150-ft radius		
Variable	Estimate	Std. error	*P*-value	Estimate	Std. error	*P*-Value	Estimate	Std. error	*P*-value
Intercept	−5.887	1.125	<0.001	−6.118	1.016	<0.001	−6.118	0.957	<0.001
Natural logarithm of major road AADT	0.528	0.124	<0.001	0.564	0.112	<0.001	0.581	0.105	<0.001
Natural logarithm of minor road AADT	0.322	0.053	<0.001	0.316	0.047	<0.001	0.305	0.044	<0.001
Number of driveways	0.135	0.158	0.390	0.118	0.073	0.104	0.075	0.037	0.044
Residential land development (base)	–	–	–	–	–	–	–	–	–
Commercial land development	0.187	0.127	0.141	0.188	0.115	0.101	0.207	0.108	0.054
Major road-2 lanes (base)	–	–	–	–	–	–	–	–	–
Major road-3 lanes	0.572	0.296	0.053	0.638	0.268	0.017	0.654	0.252	0.009
Major road-4 lanes or more	0.598	0.244	0.014	0.629	0.221	0.005	0.604	0.208	0.004
Overdispersion parameter	0.872	0.065	<0.001	0.700	0.053	<0.001	0.617	0.047	<0.001

	200-ft			250-ft			300-ft		
Variable	Estimate	Std. error	*P*-value	Estimate	Std. error	*P*-value	Estimate	Std. error	*P*-value
Intercept	−6.134	0.913	<0.001	−5.824	0.891	<0.001	−5.843	0.867	<0.001
Natural logarithm of major road AADT	0.602	0.100	<0.001	0.576	0.097	<0.001	0.578	0.095	<0.001
Natural logarithm of minor road AADT	0.292	0.041	<0.001	0.285	0.040	<0.001	0.281	0.039	<0.001
Number of driveways	0.071	0.023	0.002	0.056	0.016	<0.001	0.051	0.013	<0.001
Residential land development (base)	–	–	–	–	–	–	–	–	–
Commercial land development	0.174	0.102	0.089	0.216	0.100	0.031	0.230	0.098	0.018
Major road-2 lanes (base)	–	–	–	–	–	–	–	–	–
Major road-3 lanes	0.738	0.240	0.002	0.775	0.234	0.001	0.858	0.229	<0.001
Major road-4 lanes or more	0.619	0.198	0.002	0.633	0.193	0.001	0.715	0.189	<0.001
Overdispersion parameter	0.557	0.043	<0.001	0.533	0.041	<0.001	0.506	0.039	<0.001

	350-ft radius			400-ft radius			450-ft radius		
Variable	Estimate	Std. error	*P*-value	Estimate	Std. error	*P*-value	Estimate	Std. error	*P*-value
Intercept	−5.265	0.852	<0.001	−4.752	0.849	0.000	−4.138	0.841	<0.001
Natural logarithm of major road AADT	0.543	0.093	<0.001	0.514	0.093	0.000	0.467	0.092	<0.001
Natural logarithm of minor road AADT	0.270	0.039	<0.001	0.247	0.039	0.000	0.243	0.038	<0.001
Number of driveways	0.050	0.011	<0.001	0.048	0.010	0.000	0.048	0.009	<0.001
Residential land development (base)	–	–	–	–	–	–	–	–	–
Commercial land development	0.236	0.096	0.014	0.260	0.097	0.007	0.309	0.096	0.001
Major road-2 lanes (base)	–	–	–	–	–	–	–	–	–
Major road-3 lanes	0.748	0.224	0.001	0.838	0.224	0.000	0.763	0.221	0.001
Major road-4 lanes or more	0.607	0.185	0.001	0.648	0.185	0.000	0.564	0.182	0.002
Overdispersion parameter	0.497	0.038	<0.001	0.496	0.037	<0.001	0.502	0.037	<0.001

	500-ft			550-ft			600-ft		
Variable	Estimate	Std. error	*P*-value	Estimate	Std. error	*P*-value	Estimate	Std. error	*P*-value
Intercept	−4.030	0.821	<0.001	−4.057	0.812	<0.001	−4.030	0.800	<0.001
Natural logarithm of major road AADT	0.471	0.090	<0.001	0.486	0.089	<0.001	0.486	0.088	<0.001
Natural logarithm of minor road AADT	0.238	0.038	<0.001	0.226	0.037	<0.001	0.222	0.036	<0.001
Number of driveways	0.045	0.008	<0.001	0.042	0.007	<0.001	0.040	0.006	<0.001
Residential land development (base)	–	–	–	–	–	–	–	–	–
Commercial land development	0.326	0.094	0.001	0.332	0.093	<0.001	0.374	0.091	<0.001
Major road-2 lanes (base)	–	–	–	–	–	–	–	–	–
Major road-3 lanes	0.700	0.215	0.001	0.706	0.213	0.001	0.715	0.210	0.001
Major road-4 lanes or more	0.490	0.177	0.006	0.495	0.175	0.005	0.522	0.173	0.003
Overdispersion parameter	0.478	0.035	<0.001	0.469	0.035	<0.001	0.456	0.033	<0.001

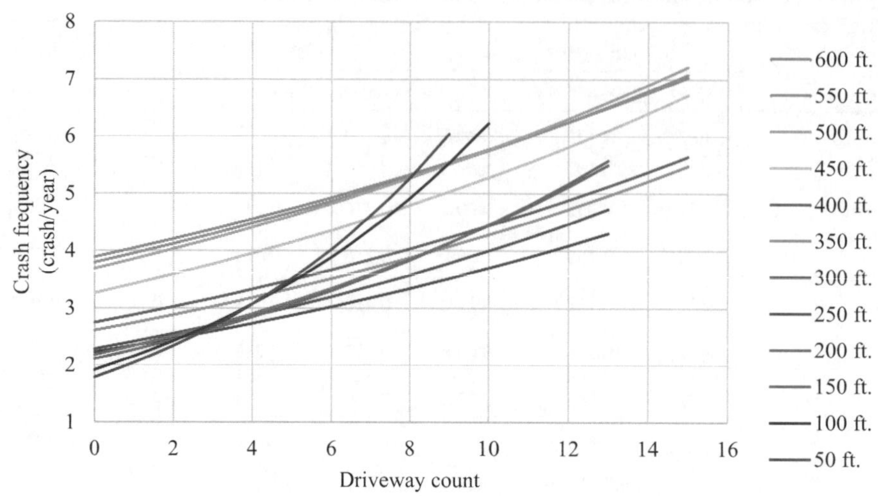

Figure 2. Expected crash frequency by driveway count by buffer radius.

effectively control the manner in which vehicles gain access to and from the public roadway network from private facilities. Based upon a review of existing guidelines from various jurisdictions, there is significant variability in policies across agencies.

In the second edition of the Access Management Manual (AMM), the upstream functional area depends on several factors, including the distance travelled during the perception-reaction time, the deceleration distance travelled while the driver slows to a stop, and the available queue storage. There are additional factors for the downstream functional area, including geometric features, operational effects, and human factors. Based on decision sight distance metrics, the AMM suggests that for a roadway in an urban area with a speed limit of 35 mph (average speed for major road in this study is 37 mph), no access points should be located within 590 ft downstream of an intersection (*18*). Additionally, a driveway information guide published by the Florida Department of Transportation (DOT) determined that the minimum distance between the intersection and driveways on roadways with a speed limit less than or equal to 45 mph should be 245 ft (*19*). Further guidance provided by the American Association of State Highway and Transportation Officials (AASHTO) suggests that if the operating speed of a roadway is 45 mph or above, the distance between the intersections and access points in an urban area should be a minimum of 300 ft (*20*). In addition to this, the Texas DOT suggested that a roadway with a speed limit of 40 mph should have a minimum corner clearance of 305 ft (*21*). Moreover, the Iowa DOT has historically required driveways to be located at least 300 ft from a primary intersection (*22*).

Ultimately, this study showed that crash rates tended to be highest when driveways were in close proximity to the intersection, particularly within a 50- to 100-ft radius. As this radius is increased, the adverse impacts of each additional driveway are mitigated to a degree, though crashes consistently increase with each access point up to the maximum 600-ft buffer. The results of this analysis provide empirical support for a minimum spacing of 300 ft for the first driveway relative to an adjacent signalized intersection. Beyond this distance, the increases in crash rates are relatively similar. At distances of 500 to 600 ft, the crash rates are nearly indistinguishable from one another.

Ultimately, this research provides important quantitative guidance in support of access management policies for location of driveways near signalized intersections. However, additional research is warranted to examine the transferability of these results to other settings. For example, the sensitivity of these results should be assessed in consideration of higher traffic volumes and speed limits. Lastly, additional information would provide further insights to the safety performance of signalized intersections when considering adjacent access points. For example, analyzing different types of crashes (i.e., rear-end crashes, right-angle crashes, etc.) would provide more focused guidance as to those crash types that are more prone to occur at specific types of driveways.

Author Contributions

The authors confirm contribution to the paper as follows: study conception and design: P. Savolainen, M. Johari, B. Bazargani, T. Barrette; data collection: T. Barrette, M. Johari, B. Bazargani; analysis and interpretation of results: P. Savolainen,

M. Johari, B. Bazargani, T. Kirsch; draft manuscript preparation: P. Savolainen, M. Johari, B. Bazargani, T. Kirsch. All authors reviewed the results and approved the final version of the manuscript.

References

1. Schultz, G. G., J. S. Lewis, and T. Boschert. Safety Impacts of Access Management Techniques in Utah. *Transportation Research Record: Journal of the Transportation Research Board*, 2007. 1994: 35–42.
2. Avelar, R. E., K. K. Dixon, L. S. Brown, M. E. Mecham, and I. van Schalkwyk. Influence of Land Use and Driveway Placement on Safety Performance of Arterial Highways. *Transportation Research Record: Journal of the Transportation Research Board*, 2013. 2398: 101–109.
3. Saxena, M. *Comparison of Various Methods To Compute Access Density and Proposing a Weighted Methodology*. MS thesis. University of South Florida, Tampa, 2009.
4. Peng, H. *Effects of Two-Way Left-Turn Lane on Roadway Safety*. MS thesis. University of South Florida, Tampa, 2004.
5. Drummond, K. P., L. A. Hoel, and J. S. Miller. *A Simulation-Based Approach to Evaluate Safety Impacts of Increased Traffic Signal Density*. No. FHWA/VTRC 02-R7. Virginia Transportation Research Council, Charlottesville, Virginia, 2002.
6. Xu, X., H. Teng, and V. Kwigizile. Safety Impact of Access Management Techniques at Signalized Intersections. *Proc., ICTIS 2011: Multimodal Approach to Sustained Transportation System Development: Information, Technology, Implementation*, American Society of Civil Engineers, Wuhan, China, 2011, pp. 403–416.
7. Oh, J., S. Washington, and K. Choi. Development of Accident Prediction Models for Rural Highway Intersections. *Transportation Research Record: Journal of the Transportation Research Board*, 2004. 1897: 18–27.
8. Xu, X., H. Teng, V. Kwigizile, and E. Mulokozi. Modeling Signalized-Intersection Safety with Corner Clearance. *Journal of Transportation Engineering*, Vol. 140, 6, 2014, p. 04014016.
9. Schultz, G., C. Allen, and T. Boschert. Making the Most of an Existing System through Access Management at Major Arterial Intersections. *Transportation Research Record: Journal of the Transportation Research Board*, 2010. 2171: 66–74.
10. Plazak, D., A. Garms, and J. Rees. Access Management Plan and Program for the Des Moines Metropolitan Area. *Proc., 2005 Mid-Continent Transportation Research Symposium*, Ames, Iowa, 2005, pp. 18–19.
11. Maze, T., and D. Plazak. *Access Management Awareness Program: Phase II Report*. Iowa Highway Research Board Project TR-402. Center for Transportation Research and Education, Iowa State University, Ames, 1997.
12. Riffkin, M., C. Allen, M. Baker, G. Schultz, and C. Kergaye. A Summary of the Economic Impacts of Raised Medians in Utah. In *Access Management Theories and Practices*, American Society of Civil Engineers, Shanghai, China, 2015, pp. 12–22. https://ascelibrary.org/doi/book/10.1061/9780784413869.
13. Benz, R. J., N. Norboge, A. Voigt, and S. Gage. Economic Assessment of Access Management Projects in the Houston, Texas, Region. *Transportation Research Record: Journal of the Transportation Research Board*, 2015. 2486: 80–89.
14. Chin, H. C., and M. A. Quddus. Applying the Random Effect Negative Binomial Model to Examine Traffic Accident Occurrence at Signalized Intersections. *Accident Analysis & Prevention*, Vol. 35 No. 2, 2003, pp. 253–259.
15. Wang, X., Y. Song, R. Yu, and G. G. Schultz. Safety Modelling of Suburban Arterials in Shanghai, China. *Accident Analysis & Prevention*, Vol. 70, 2014, pp. 215–224.
16. Dixon, K. K., and R. E. Avelar. Validation Technique Applied to Oregon Safety Performance Function Arterial Segment Models. *Transportation Research Record: Journal of the Transportation Research Board*, 2015. 2515: 115–123.
17. Bahar, G., and E. Hauer. *NCHRP Project 20-07/Task 332: User's Guide to Develop Highway Safety Manual Safety Performance Function Calibration Factors*. Transportation Research Board of the National Academies, Washington, D.C., 2014.
18. Williams, K. M., V. G. Stover, K. K. Dixon, and P. Demosthenes. *Access Management Manual*, 2nd ed. Transportation Research Board of the National Academics, Washington D.C., 2014.
19. Sokolow, G., V. Stover, F. Broen, and A. Dartz. *Driveway Information Guide*. Florida Department of Transportation, Tallahassee, Florida, 2008. http://www.fdot.gov/planning/systems/programs/sm/accman/pdfs/driveway2008.pdf.
20. AASHTO. *Policy on Geometric Design of Highways and Streets*. AASHTO, Washington, D.C., 2001.
21. Frawley, W. E., and W. L. Eisele. *Access Management Guidebook for Texas*. Texas Transportation Institute, Texas A&M University System, College Station, Texzs, 2005.
22. Iowa Department of Transportation. *Iowa Primary Highway Access Management Policy*. Iowa Department of Transportation, Ames, Iowa, 2012.

The Standing Committee on Access Management (AHB70) peer-reviewed this paper (18-00010).

TRR
JOURNAL OF THE TRANSPORTATION RESEARCH BOARD

Article

Transportation Research Record
2018, Vol. 2672(17) 22–34

DOI: 10.1177/0361198118787978
journals.sagepub.com/home/trr
SAGE

A Capacity Estimation Model for a Contraflow Left-Turn Pocket Lane at Signalized Intersections

Yi Zhao[1], Rachel M. James[1,2], Lin Xiao[1], and Joe Bared[1]

Abstract

Alternative intersection designs are increasingly proposed and adopted by different agencies to meet the needs of growing traffic demand and constrained transportation resources. The left turn (LT) is one of the most critical movements at signalized intersections from both a safety and operations perspective. Heavy LT volumes are especially impactful to the operational efficiency of a signalized intersection and often result in queue spillback. A contraflow left-turn pocket lane (CLPL) is proposed to mitigate congestion caused by heavy LT demand and has been shown in simulation to greatly mitigate the impact of queue spillback. The CLPL dynamically uses the opposing through lane (OTL) as an additional LT lane within the signal cycle on a temporary basis when the OTL is not occupied by through traffic. While geometric design schematics and analytical procedures for estimating delay have been proposed and discussed in existing literature, methodologies for estimating capacity benefits and traffic operations are not yet well defined. This paper has three primary contributions to the literature: development of a probabilistic capacity estimation model, exploration of the impact of key characteristics (e.g., cycle length, LT demand, lane selection preference) on estimated intersection capacity, and recommendations for the real-world implementation of a CLPL. The simulation results indicate that the CLPL treatment can increase a signalized intersection's throughput up to 25% and decrease the intersection's average delay by 35%.

Left turns are one of the most critical maneuvers at signalized intersections from both a safety and operations perspective (*1*). A 2010 U.S. Department of Transportation study analyzing crash factors found that the most common pre-crash event at intersections was a vehicle attempting to make a left turn (LT) (*2*); additionally, 17% to 32% of pedestrian crashes at intersections can be attributed to LT vehicles (*3*). The most common countermeasure to a high LT volume and a high number of right-angle crashes is the implementation of protected or protected/permissive phasing.

A high percentage of LT vehicles can also negatively affect the efficiency of a signalized intersection if the turn bay is insufficient and queue spillback occurs (*4*). Queue spillback can significantly reduce the effective capacity of a signal because it limits the ability of a lane to serve through demand (*5*, *6*). Haddad and Geroliminis (*6*) found that a "significant" reduction in intersection capacity is observed as a result of queue spillback. However, the negative impact of queue spillback is not limited to the isolated intersection where it occurs: it has far reaching effects along the downstream corridor by decreasing bandwidth and negating the positive impacts of progression and platooning (*4*, *6*). A model for signalized arterial travel time estimation capable of capturing the effect of queue spillback on link capacity found that the occurrence of link spillover can increase travel times by 50–100% for short distance links (*7*). Current best practices for mitigating the impact of queue spillover include longer turn bays and leading protected–permissive LT phasing (*1*, *8*). Thus, the development of signalized intersection treatments that maximize the number of vehicles serviced during LT protected phasing and minimize the likelihood that queue spillback will occur have the potential to greatly improve traffic flow at corridors with signalized intersections.

This paper presents a probabilistic capacity estimation model for an innovative intersection treatment called a contraflow left-turn pocket lane (CLPL), which increases LT capacity without increasing the footprint of the

[1]Turner-Fairbank Highway Research Center, Federal Highway Administration, McLean, VA
[2]The University of Texas at Austin, Austin, TX

Corresponding Author:
Address correspondence to Rachel M. James: rachel.james@dot.gov

intersection. There are four requirements for the implementation of this treatment: (i) leading protected LT phases on both streets, (ii) prohibited permissive LTs on either street, (iii) cross-streets have at least two receiving lanes, and (iv) right turns from the cross-streets are funneled away from the CLPL to avoid conflicts with queueing traffic. An illustration of a signalized intersection with a CLPL treatment is shown in Figure 1. The operation can be summarized as follows: when the approach's opposing through lane (OTL) is not in use by LT vehicles from the cross-street, a pre-signal indicates to the LT vehicles on the approach that they may safely queue in the traditional LT bay or in the CLPL; this essentially creates an additional LT bay without increasing the surface area of the intersection. When the protected LT phase for that approach obtains priority at the main signal, both LT bays are able to discharge into separate receiving lanes on the cross-street. Before the protected LT phase ends, the pre-signal terminates to give vehicles in the OTL time to discharge before the protected LT phase ends at the main signal. Though none are currently operational in North America, there appear to be over 50 successful implementations in China (*9*).

This intersection intervention strategy increases the number of vehicles serviced during the valuable protected LT phase in a cycle and has been shown to be an effective queue spillback mitigation strategy via simulation. This strategy can potentially be a low-cost solution because it requires no additional right-of-way (ROW) or major reconstruction of existing intersections (*10*).

The remainder of this paper is organized as follows: (1) Literature related to contraflow (reversible-flow) operations is detailed, specifically focusing on existing research on CLPLs. (2) A stochastic model is built to estimate the capacity of the CLPL based on vehicle arrival patterns and lane selection behavior. (3) Numerical analyses are detailed, using both the proposed analytical probabilistic model and Vissim simulations. (4) Efforts

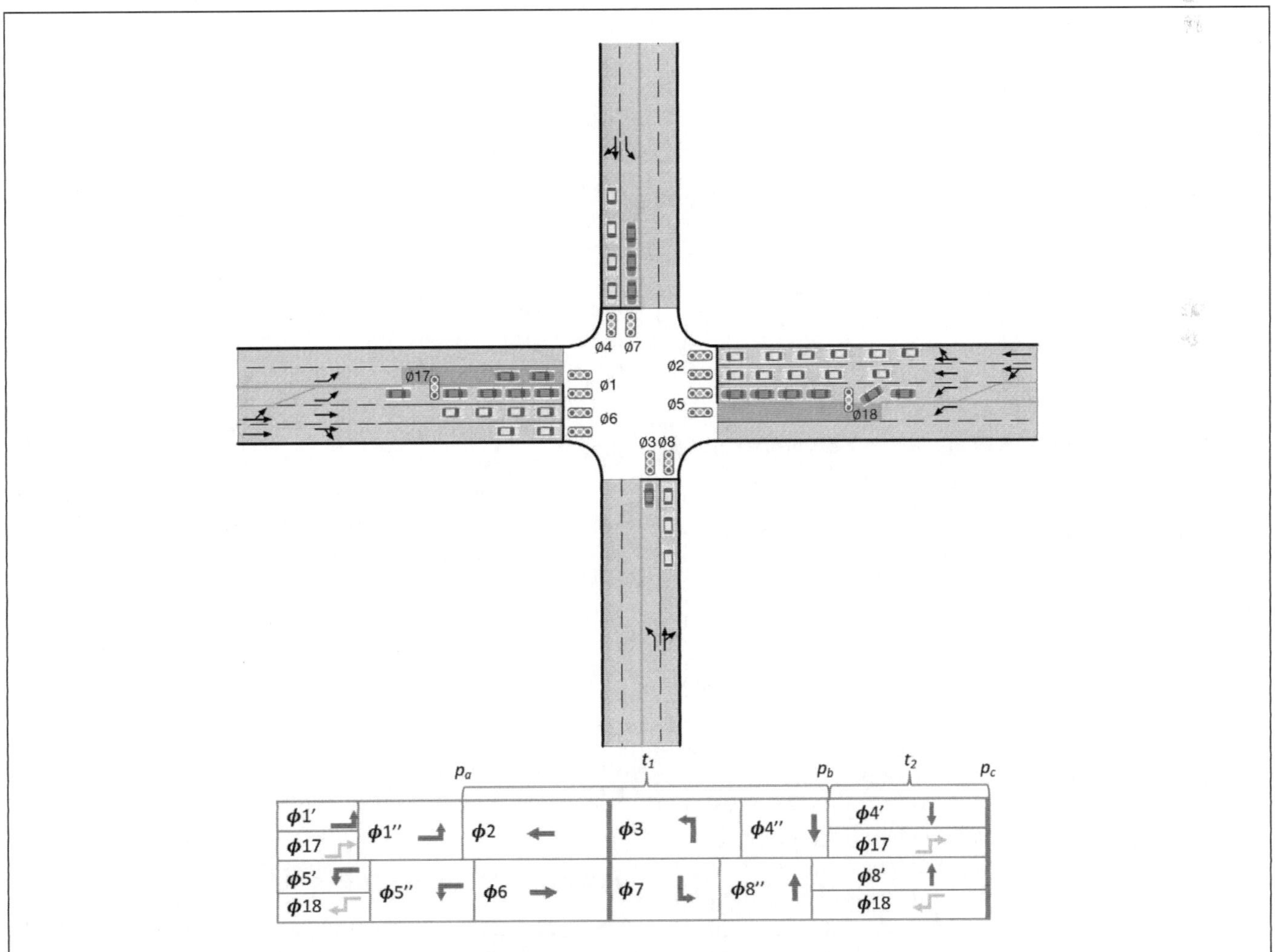

Figure 1. CLPL signal timing plan with critical time points identified and corresponding intersection geometry.

to implement CLPL in practice are detailed, including updates on an in-progress research project at the FHWA Turner-Fairbank Highway Research Center (TFHRC) to develop signing and marking strategies to facilitate real-world deployment. Finally, there is a summary of conclusions and recommendations.

Literature Review

Contraflow (reversible-flow) lanes are not a new concept; they have been used in the United States for more than 80 years (*11*). There are six common applications of contraflow lanes in the U.S.; these applications include, but are not limited to, congestion mitigation and planned special events (*12*). Reversible lane operations have been deployed in nearly every major city, and in a growing number of small- and medium-sized cities, because of their cost effectiveness.

Contraflow lanes are popular for their ability to mitigate the directionality problem of traffic flow without the cost or environmental impact associated with the construction of new lanes. According to the *Traffic Engineering Handbook*, contraflow lanes are a proven technique for increasing peak-hour capacity of existing streets under certain conditions (*13*). Contraflow lanes have been found to work best when demand volumes are near capacity, congestion patterns are predictable, ROW is limited, 2:1 ratios of primary-to-secondary direction volume exist, and when there is inadequate capacity on adjacent streets (*14*). Additionally, despite apprehension from planning agencies that reversible operations are "dangerous," "complicated," and "controversial," most research indicates that drivers readily adapt to reversible operations with relative ease (*12*, *14*).

An NCHRP synthesis of reversible lanes in practice identified a substantial gap in knowledge surrounding the real-world performance of reversible lanes from a quantitative perspective: before and after studies examining safety, travel time, and traffic volume are not readily available in the literature. Thus, practical benefits of reversible traffic streams are not known with any degree of certainty (*14*). The general consensus is that the lane utilization ratio of a reversible lane tends to be lower than that of a normal lane (*15*); the synthesis of reversible lanes reports that the capacity of a contraflow freeway lane ranges from 0.5 to 1 times the capacity of an existing normal-flow lane (*14*). Lambert and Wolshon conducted a follow-up project analyzing two contraflow lanes: one reversed for special event traffic and the other reversed during peak-hour (*16*). They found that when the facility is experiencing moderate flow, the throughput of the contraflow lanes is lower than that of a normal-flow lane at a statistically significant level. However, as the volume approaches capacity of the corridor, the volume and speed of the contraflow lanes approaches that of the normal-flow lane (*16*).

CLPLs, originally referred to as exit lanes for left turns (EFL), were introduced as a queue mitigation strategy at signalized intersections in 2013 (*17*). This work sought to maximize the reserve capacity of the "mixed usage area" as a function of geometric layout, intersection signal timing, and pre-signal timing. This project found that under heavy LT volumes, delay was reduced up to 50% and queue length was reduced up to 73%; for each additional approach using CLPL there was about a 7% increase in capacity (*17*). Krause et al. (*18*) built on this cornerstone work to create an analytical method that calculates the clearance time based on intersection geometrics and the estimated 95th percentile of the queue length. A real-world case study in Rockville, Maryland, revealed possible intersection delay reductions up to 22%. The most recent article published in this space was written by Wu et al. (*19*). This work proposed an analytical delay model for estimating the delay to LT vehicles at intersections with the CLPL design. Additionally, they created a capacity estimation model for isolated intersections. The model was calibrated and validated using field data from six operational CLPLs in Handan, China. They found that the CLPL treatment resulted in less control delay for both the LT and through vehicles.

One limitation of existing capacity and delay estimation models for CLPLs is that they do not include the stochastic nature of vehicle arrival patterns in the capacity analysis. This is a critical component of signalized intersection modeling, as existing research on regular LT lanes has shown the importance of stochastic analysis to account for capacity loss in certain situations. Tian et al. (*20*) found that capacity gains from a regular LT lane is stochastic and a function of the randomness of vehicle arrival. Wu (*21*) proposed a framework for using simulation data to calibrate a theoretical-empirical model, where the total approach capacity with shared/exclusive turning lanes can be estimated. Kikuchi et al. (*22*, *23*) developed a general framework, accounting for the stochastic nature of vehicle arrival, to calculate the required length of a LT bay that prevents LT queue spillback and blockage of the LT bay by queueing vehicles in the adjacent lane. This research seeks to fill a similar void in the contraflow literature by adopting the concept of random vehicle arrival and drivers' stochastic lane selection behavior to create a capacity model to provide more accurate CLPL treatment capacity estimation results.

Proposed Capacity Model

This section details the proposed stochastic model for estimating the capacity of a planned CLPL. There is a discussion of the enumeration of nine possible queueing

Table 1. Notation Used in Development of Capacity Model

$c_{\phi x}$	capacity of the phase x
$c^{J}_{s_k}$	capacity "loss" in scenario k relaxing the assumption of perfect use of CLPL by drivers
C	cycle length
i	number of LT vehicles arrived during the t_1; $i \in \{0, 1, 2, \ldots\}$
j	number of LT vehicles arrived during the t_2; $j \in \{0, 1, 2, \ldots\}$
k	scenario identifier; $k \in \{1, 2, \ldots 9\}$
L_{CLP}	length of the CLPL (feet)
n	number of CLPL per approach
N	maximum number of vehicles that can be stored in the CLPL
s_k	LT vehicle queueing scenario k
$s_{\phi x}$	saturation flow rate of the phase x
p	the probability of drivers choosing RLPL. Therefore, $1 - p$ is the probability of drivers choosing CLPL
q_L	total traffic volume for LT vehicles (vph)
R_{t1}, R_{t2}	red time for LT vehicle during t_1 and t_2, respectively (s)
v	design speed for CLPL (mph)
$\lambda_{t1}, \lambda_{t2}$	average LT vehicle arrival that queueing during t_1 and t_2 time, respectively, $\lambda_{t1} = \frac{q_L R_{t1}}{3600}, \lambda_{t2} = \frac{q_L R_{t2}}{3600}$
t_1	the time that queueing occurs in the RLPL while the CLPL is closed
t_2	the time that queueing occurs while the CLPL is open
$P(i)$	probability of i vehicles arriving during t_1
$P(j)$	probability of j vehicles arriving during t_2
$P(i,j)$	conditional probability of i and j vehicles arriving during t_1 and t_2, respectively
$E(c^{J}_{k})$	expected capacity loss of the kth scenario, $\forall k = 1, 2, \ldots, 9$
r	total number of people that selected the CLPL who arrived in t_2
$P(r\|(i,j))$	probability of j people arriving t_2 and choosing to take the CLPL (r) given the conditional probability that i people arrive during t_1 and j people arrive during t_2

scenarios by unique assumptions of arrival patterns. For each of the nine possible scenarios, the model calculates the expected capacity "loss" due to suboptimal use of the pocket lane; this is described later. This capacity loss is applied to insights derived from the *Highway Capacity Manual* to determine the expected change in capacity as a result of the implementation of the CLPL. The notation used in this section is summarized in Table 1.

Possible Queueing Scenarios for LT Vehicles

The capacity "loss" from the maximum possible increase in capacity attributable to the introduction of a new lane is calculated first. To calculate the capacity loss for the CLPL due to drivers' suboptimal use of the additional pocket, LT queue length at both the regular left-turn pocket lane (RLPL) and CLPL should be calculated at two critical time points, t_1 and t_2, as depicted in Figure 1. Note that Phase 1 corresponds to the dedicated LT phase and Phase 17 is the pre-signal time that allows for safe queueing in the CLPL. Figure 1 also shows how each of the phases correspond with movements at the intersection in question.

At the critical time points t_1 and t_2, a possible queue pattern can be categorized into one of nine enumerated vehicle arrival scenarios, which is denoted as S_k where $k = 1, 2, \ldots, 9$. They are summarized in the bulleted list below and in Figure 2. Note that light grey cars indicate vehicles arriving during t_1, while the darker grey cars are the vehicles arriving during t_2.

$S_1 - S_4$: In these four scenarios, LT vehicle arrival at the end of t_1 is less than the maximum number of vehicles (N) that can be stored in the CLPL.

- S_1: Total arrivals (during t_1 and t_2) are $< N\text{-}1 \mid$ neither RLPL and CLPL are fully utilized, capacity loss due to low LT vehicle arrival (no blockage).
- S_2: Total arrivals are $< 2N\text{-}1 \mid$ capacity loss occurs because CLPL is blocked by LT vehicles that arrived in t_2.
- S_3: Total arrivals are $< 2N\text{-}1 \mid$ capacity loss occurs because CLPL is not fully utilized (though no blockage).
- S_4: Total arrivals are $\geq 2N\text{-}1 \mid$ capacity loss occurs because CLPL is blocked by LT vehicle that arrived in t_2.

$S_5 - S_8$: In these four scenarios, LT vehicle arrival at the end of t_1 is greater than or equal to the maximum number of vehicles (N) and less than $2N\text{-}1$.

- S_5: Total arrivals are $< 2N\text{-}1 \mid$ CLPL is blocked by LT vehicle that arrived in t_1.
- S_6: Total arrivals are $< 2N\text{-}1 \mid$ CLPL is not fully utilized (no blockage).

S_p	Vehicle Queueing at t_1	Vehicle Queueing at t_2	Probability Function $P(S_k)$	Capacity Loss c^l
1	$i \le N\text{-}1$	$i \le i+j \le N\text{-}1$	$\sum_{i=0}^{N-1}\frac{\lambda_{t1}^{i}e^{-\lambda_{t1}}}{i!}\sum_{j=0}^{N-1-i}\frac{\lambda_{t2}^{j}e^{-\lambda_{t2}}}{j!}\sum_{r=0}^{j}C_r^j(1-p)^r p^{j-r}$	$N\text{-}r$
2	$i \le N\text{-}1$	$N\text{-}1 < i+j \le 2N\text{-}2$	$\sum_{i=0}^{N-1}\frac{\lambda_{t1}^{i}e^{-\lambda_{t1}}}{i!}\sum_{j=0}^{2N-2-i}\frac{\lambda_{t2}^{j}e^{-\lambda_{t2}}}{j!}\sum_{r=0}^{j-(N-1-i)-1}C_r^{N-1-i+r}(1-p)^r p^{N-1-i}p$	$N\text{-}r$
3	$i \le N\text{-}1$	$N\text{-}1 < i+j \le 2N\text{-}2$	$\sum_{i=0}^{N-1}\frac{\lambda_{t1}^{i}e^{-\lambda_{t1}}}{i!}\sum_{j=N-i}^{2N-2-i}\frac{\lambda_{t2}^{j}e^{-\lambda_{t2}}}{j!}\sum_{r=j-(N-1-i)}^{\min(N-1,j)}C_r^j(1-p)^r p^{j-r}$	$N\text{-}r$
4	$i \le N\text{-}1$	$2N\text{-}1 \le i+j$	$\sum_{i=0}^{N-1}\frac{\lambda_{t1}^{i}e^{-\lambda_{t1}}}{i!}\sum_{j=2N-1-i}^{\infty}\frac{\lambda_{t2}^{j}e^{-\lambda_{t2}}}{j!}\sum_{r=0}^{N-1}C_r^{N-1-i+r}(1-p)^r p^{N-1-i}p$	$N\text{-}r$
5	$N\text{-}1 < i \le 2N\text{-}2$	$0 \le j$	$\sum_{i=N}^{2N-2}\frac{\lambda_{t1}^{i}e^{-\lambda_{t1}}}{i!}\sum_{r=0}^{i-(N-1)-1}(1-p)^r p$	$N\text{-}r$
6	$N\text{-}1 < i \le 2N\text{-}2$	$i \le i+j \le 2N\text{-}2$	$\sum_{i=N}^{2N-2}\frac{\lambda_{t1}^{i}e^{-\lambda_{t1}}}{i!}\sum_{j=0}^{2N-2-i}\frac{\lambda_{t2}^{j}e^{-\lambda_{t2}}}{j!}(1-p)^{i-(N-1)}(1-p)^j$	$N\text{-}[i\text{-}(N\text{-}1)]\text{-}j$
7	$N\text{-}1 < i \le 2N\text{-}2$	$i \le i+j \le 2N\text{-}2$	$\sum_{i=N}^{2N-2}\frac{\lambda_{t1}^{i}e^{-\lambda_{t1}}}{i!}\sum_{j=0}^{2N-2-i}\frac{\lambda_{t2}^{j}e^{-\lambda_{t2}}}{j!}(1-p)^{i-(N-1)}\sum_{r=0}^{j-1}(1-p)^r p$	$N\text{-}[i\text{-}(N\text{-}1)]\text{-}r$
8	$N\text{-}1 < i \le 2N\text{-}2$	$2N\text{-}1 \le i+j$	$\sum_{i=N}^{2N-2}\frac{\lambda_{t1}^{i}e^{-\lambda_{t1}}}{i!}\sum_{j=2N-1-i}^{\infty}\frac{\lambda_{t2}^{j}e^{-\lambda_{t2}}}{j!}(1-p)^{i-(N-1)}\sum_{r=0}^{2N-2-i}(1-p)^r p$	$N\text{-}[i\text{-}(N\text{-}1)]\text{-}r$
9	$2N\text{-}1 \le i$	$0 \le j$	$\sum_{i=2N-1}^{\infty}\frac{\lambda_{t1}^{i}e^{-\lambda_{t1}}}{i!}\sum_{r=0}^{N-1}(1-p)^r p$	$N\text{-}r$

Figure 2. Scenarios of LT queueing at different critical time points.

- S_7: Total arrivals are $<2N\text{-}1$ | CLPL is blocked by LT vehicle that arrived in t_2.
- S_8: Total arrivals are $\ge 2N\text{-}1$ | CLPL is blocked by LT vehicle that arrived in t_2.

S_9: LT vehicle arrival at the end of t_1 is $\ge 2N\text{-}1$, which indicates that CLPL can be either filled up or blocked by the vehicle that arrived in t_1.

Probabilistic Model for Capacity "Loss" of CLPL Due to Suboptimal Use

The previous section enumerated the possible cases in which the CLPL may be used in a suboptimal manner due to lane blockages or underused space due to driver preference. This section derives the probabilistic functions for expected capacity loss for each of the nine scenarios, using Scenario 1 as an example. This section closes with an explanation of how to leverage the capacity loss of each scenario to calculate the capacity loss of the CLPL due to suboptimal use (compared with a fully utilized RLPL operating at capacity).

Kikuchi (*22*) suggested that vehicle arrival is random and follows a Poisson distribution. The probability of i,j vehicles arriving during t_1 and t_2, respectively, can be denoted by:

$$P(i) = \frac{\lambda_{t_1}^{i}e^{-\lambda_{t_1}}}{i!} \tag{1}$$

$$P(j) = \frac{\lambda_{t_2}^{j}e^{-\lambda_{t_2}}}{j!} \tag{2}$$

The conditional probability of i and j LT vehicles arriving during critical time points t_1 and t_2, respectively, can be denoted by:

$$P(i,j) = \frac{\lambda_{t_1}^{i}e^{-\lambda_{t_1}}}{i!}\frac{\lambda_{t_2}^{j}e^{-\lambda_{t_2}}}{j!} \tag{3}$$

Assume that during the t_1 time period i total vehicles arrive at the intersection. All LT vehicles arriving during t_1 can only select RLPL, because CLPL is closed during the t_1 period. However, taking Scenario 1 from Figure 2 as an example, if j vehicles arrive during the t_2 period, then each vehicle can choose to queue in the RLPL or CLPL. If r denotes the total number of vehicles selecting CLPL who arrived during t_2 period, then r follows a negative binomial distribution and the probability function is given by the following equation:

$$P(r) = C_r^j(1-p)^r p^{j-r} = \binom{j}{r}(1-p)^r p^{j-r} \quad (4)$$

The expectation of the capacity loss $E(c_1^l)$ in Scenario 1 is calculated using the following equation:

$$E(c_1^l) = \sum P(s_1)c_1^l = \sum\sum\sum P(r|(i,j))c_1^l = \sum_{i=0}^{N-1}\sum_{j=0}^{N-1-i}\sum_{r=0}^{j}\frac{\lambda_{t_1}^i e^{-\lambda_{t_1}}}{i!}\frac{\lambda_{t_2}^j e^{-\lambda_{t_2}}}{j!}C_r^j(1-p)^r p^{j-r}(N-r) \quad (5)$$

The LT arrival rate in t_1 will affect the future arriving vehicles' decision regarding the CLPL based on the arrival pattern in t_1. Scenarios 1–4, Scenarios 5–8, and Scenario 9 are categorized as three different groups to further calculate the expectation of the capacity loss:

$$\begin{aligned} E(c_{1-4}^l) &= \sum P(i \le N-1)c^l \\ &= \sum_{i=0}^{N-1}\frac{\lambda^i e^{-\lambda}}{i!}\left\{\sum_{j=0}^{N-1-i}\frac{\lambda^j e^{-\lambda}}{j!}\sum_{r=0}^{j}C_r^j(1-p)^r p^{j-r}(N-r)\right. \\ &+ \sum_{j=N-i}^{2N-2-i}\frac{\lambda^j e^{-\lambda}}{j!}\left[\sum_{r=0}^{j-(N-1-i)-1}C_r^{N-1-i+r}(1-p)^r p^{N-1-i}p(N-r) + \sum_{r=j-(N-1-i)}^{\min(N-1,j)}C_r^j(1-p)^r p^{j-r}(N-r)\right] \\ &+ \left.\sum_{j=2N-1-i}^{\infty}\frac{\lambda^j e^{-\lambda}}{j!}\sum_{r=0}^{N-1}C_r^{N-1-i+r}(1-p)^r p^{N-1-i}p(N-r)\right\} \end{aligned} \quad (6)$$

$$\begin{aligned} E(c_{5-8}^l) &= \sum P(N-1<i\le 2N-2)c^l \\ &= \sum_{i=N}^{2N-2}\frac{\lambda^i e^{-\lambda}}{i!}\left\{\sum_{r=0}^{i-(N-1)-1}(1-p)^r p(N-r)\right. \\ &+ \sum_{j=0}^{2N-2-i}\frac{\lambda^j e^{-\lambda}}{j!}\left[(1-p)^{i-(N-1)}(1-p)^j(N-(i-(N-1))-j)\right. \\ &+ \left.(1-p)^{i-(N-1)}\sum_{r=0}^{j-1}(1-p)^r p(N-(i-(N-1))-r)\right] \\ &+ \left.\sum_{j=2N-1-i}^{\infty}\frac{\lambda^j e^{-\lambda}}{j!}(1-p)^{i-(N-1)}\sum_{r=0}^{2N-2-i}(1-p)^r p(N-(i-(N-1))-r)\right\} \end{aligned} \quad (7)$$

$$E(c_9^l) = \sum P(i>2N-1)c_9^l = \sum_{i=2N-1}^{\infty}\frac{\lambda^i e^{-\lambda}}{i!}\sum_{r=0}^{N-1}(1-p)^r p(N-r) \quad (8)$$

The capacity loss for the entire pocket attributed to the nine aforementioned scenarios of lane blockage or underused CLPL space can be further expressed as the summation of the expectation of the capacity loss for each scenario enumerated above, as in (9):

$$c_t^l = \sum_{i=1}^{9}E(c_i^l) \quad (9)$$

Stochastic Model for Expected Capacity of a CLPL

In the previous section, the capacity loss of the CLPL, accounting for the possible scenarios of lane blockage or underused pocket lane space, is derived. This section derives the stochastic model for the total expected capacity of a CLPL via insights obtained from the 2010 Edition of the *Highway Capacity Manual* (HCM) (5). The realistic capacity gain of the CLPL is limited by either the theoretical capacity of the protected LT phase (ø1), which accounts for the capacity loss derived in the previous section, or the theoretical capacity of the pre-signal (ø17), which accounts for the appropriate losses in capacity due to the necessary clearance time of the CLPL before phase ø1 ends.

Table 2. Simulation Settings

Parameters	Value
Simulation time horizon (s)	3,600
Warm-up time (s)	900
Signal control strategy	Semi-actuated
Saturation flow rate (vph)	1,830
Maximum speed (km/h)	68
Desired speed (km/h)	60
Maximum acceleration (m/s^2)	3.5
Desired acceleration (m/s^2)	1.37
Maximum deceleration (m/s^2)	7.9
Desired deceleration (m/s^2)	3.0
Average headway distance (m)	2.0
Minimum headway distance (m)	1.5
Minimum headway (s)	0.9

Theoretical Capacity for LT Phase ø1. The capacity model for a protected LT lane given in the HCM (*5*) is:

$$c_{\phi 1} = \frac{g_{\phi 1}}{C} s_{\phi 1} \tag{10}$$

where $c_{\phi 1}$ represents the maximum possible capacity gain of the CLPL, with a phase length of $g_{\phi 1}$, a saturation flow rate of $s_{\phi 1}$, and the assumption that no capacity loss from suboptimal use occurs. The second term in Equation 11 accounts for the relaxation of the assumption of perfect use, as derived earlier via the nine scenarios of possible capacity loss.

$$c_{\phi 1} = \frac{g_{\phi 1}}{C} s_{\phi 1} - \frac{3600}{C}\left(E\left(c^l\right)\right) \tag{11}$$

Theoretical Capacity for Pre-Opening Phase ø17. The split of the phase (ø17) is associated with two regular phases: major street LT phase (ø1) and cross-street LT phase (ø4). To ensure the operational safety, dummy phases ø1" and ø4" are introduced to make sure no vehicle conflicts occur in the CLPL from cross-street turning vehicles (see Figure 1) while queueing is allowed in the CLPL. Therefore, the theoretical capacity of phase ø17 can be calculated as:

$$\begin{aligned} c_{\phi 17} &= \frac{g_{\phi 1} + g_{\phi 4} - g_{\phi 1''} - g_{\phi 4''}}{C} s_{\phi 17} \\ &= \frac{g_{\phi 1} + g_{\phi 4} - 2 * \frac{L_{CLP}}{v} * 0.6818}{C} s_{\phi 17} \end{aligned} \tag{12}$$

Theoretical Capacity for Entire LT Approach. The capacity of the CLPL design is restricted by both the capacity of the CLPL and the capacity of the pre-opening entrance. Therefore, the final capacity of the CLPL design can be calculated as the minimum of the following two parts:

$$c = min\left\{c_{\phi 1}, c_{\phi 17}\right\} \tag{13}$$

Numeric and Simulation Analysis

The proposed statistical analytical model is evaluated through a series of microscopic simulations completed in Vissim because there are no existing implementations of CLPL intersections in the U.S. The selected measures of effectiveness (MOEs) for this study are throughput and total delay of both the LT approach and the entire intersection. The test site is a hypothetical intersection, illustrated in Figure 1. The eastbound/westbound direction is designed as the major street and is where the CLPL is implemented. The northbound/southbound direction is designed as the cross-street.

The signal is configured with semi-actuated control; this means all the signal timing for LT and cross-street through traffic is actuated based on the vehicle arrival; unused green phase is accumulated and reassigned to the through phase on the major street. Because the cycle length for the fully-actuated control strategy is uncertain, the semi-actuated control strategy is adopted to ensure all signals along the major street corridor can maintain coordination.

The simulation model has been calibrated based on the default saturation flow rate provided by HCM 2010 (*5*). Detailed simulation settings can be found in Table 2.

Simulation Analysis

A multitude of variables were incorporated into this analysis to better understand their impact on CLPL performance; these variables included demand level, cycle length, opening position, and balanced/unbalanced demand. There were six different simulations representing varying LT demand, ranging from 200 to 700 vph in intervals of 100 vph. The cycle length ranged from 60 to 150 s in intervals of 30 s for a total of four simulations. The pocket opening position varied from 150 to 250 ft in intervals of 50 ft for a total of three simulations. Additionally, both balanced and unbalanced demand simulations were evaluated. Thus, there were 144 different combinations of the aforementioned variables to explore CLPL performance. Ten random seeds were run for each combination of the aforementioned variables to obtain more statistically reliable results, resulting in a grand total of 1,440 simulations.

CLPL Performance as a Function of Various Demand Levels. As Figure 3 indicates, the CLPL design was shown to increase the LT vehicle throughput by 65% and decrease the LT vehicle delay by 55%. The CLPL design also benefits the adjacent through vehicles. The detailed simulation results indicate that at the intersection-level the CLPL design can effectively reduce the capacity drop during the high-demand peak-hour scenario. As

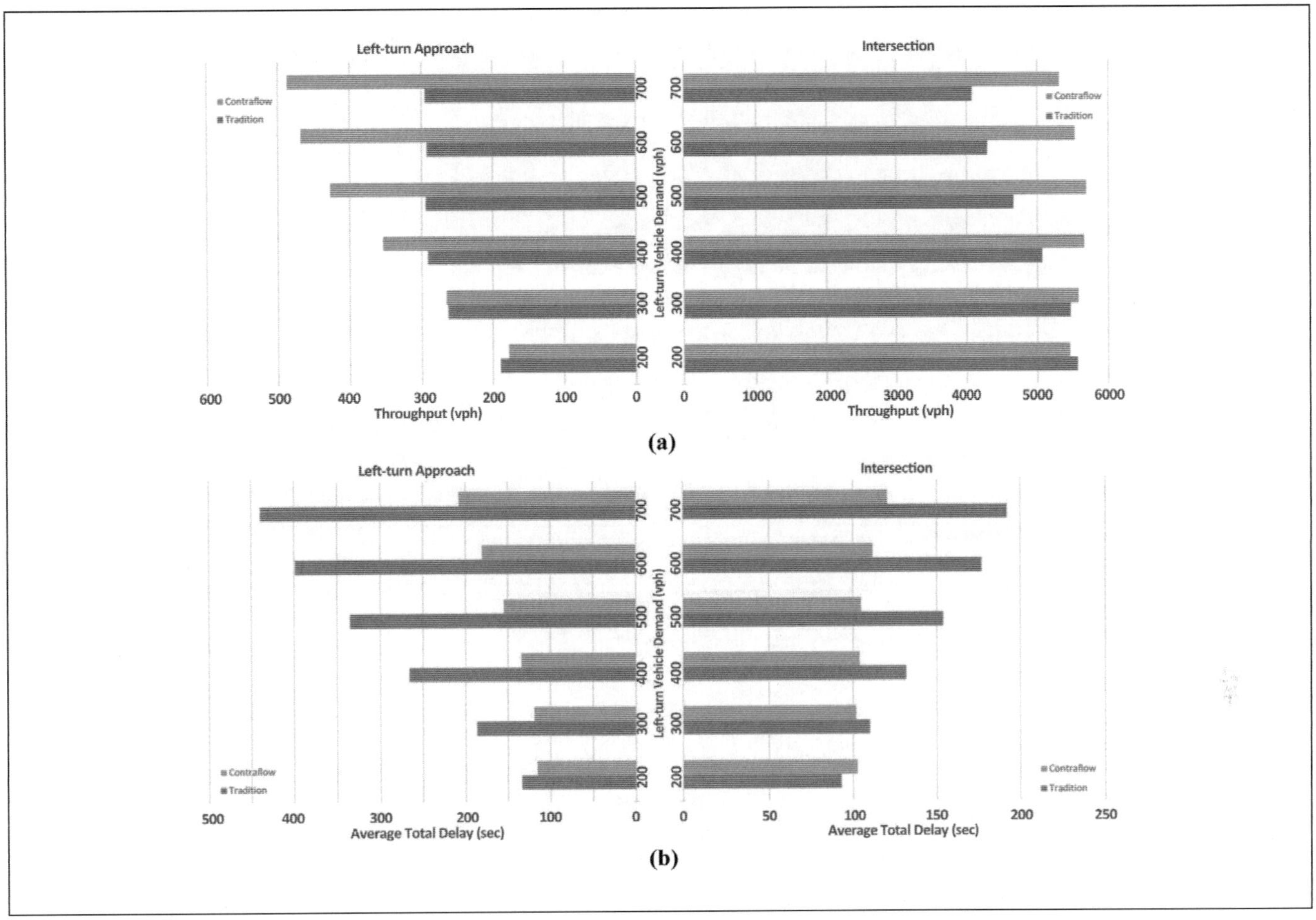

Figure 3. MOEs as a function of varying demand levels: (*a*) throughput comparison for LT approach and intersection and (*b*) delay comparison for LT approach and intersection.

Figure 3*a* illustrates, the throughput for the traditional design dropped by 26%, while the CLPL design only dropped by 3% (both under congested conditions).

CLPL Performance as a Function of Various Cycle Lengths. The performance of the CLPL as a function of different cycle lengths is plotted in Figure 4. Figure 4*a* shows the change in throughput as a function of cycle length, while Figure 4*b* shows the change in delay as a function of cycle length. The two figures on the left represent the impact on the LT approach only, while the two figures on the right represent the impact on the entire intersection.

The CLPL performs better for shorter cycle lengths under the same LT demand level, which is a reasonable finding as the v/c ratio is negatively correlated with cycle length (see Figure 4). The maximum absolute capacity enhancement is achieved at 90 s and 120 s cycle length scenarios of different demand levels. In summary, in order to maximize the throughput while achieving acceptable intersection-level delay, cycle lengths ranging from 90 to 120 s are recommended for actual operation.

CLPL Performance as a Function of Shifting Protected LT Green to Through Traffic. The analyses strongly indicate that the CLPL treatment can greatly reduce the delay for LT traffic; therefore, an interesting secondary question relates to the potential benefit of shifting the saved protected LT green split to through traffic. This scenario is explored via simulation with a v/c ratio of roughly 1.0 (LT demand at 300 vph). Three different "time-shifting" simulations (i.e., runs where 0 s, 3 s, and 6 s were shifted from the LT phase to the through phase) were analyzed under different through traffic demand schemes (ranging from 900 vph to 1500 vph, under 100 vph increments).

The results are demonstrated in Figure 5. The bar chart represents the absolute value reported by the simulation and the line chart indicates the relative change for 3 s and 6 s of green time shifted from the LT phase to the through phase. Figure 5*a* indicates that for every 3 s shifted from the LT phase to the through phase, the CLPL design can increase the intersection's capacity by up to 5% (roughly 150 vph). This translates to a 30% to 45% reduction in delay per vehicle. However, after the

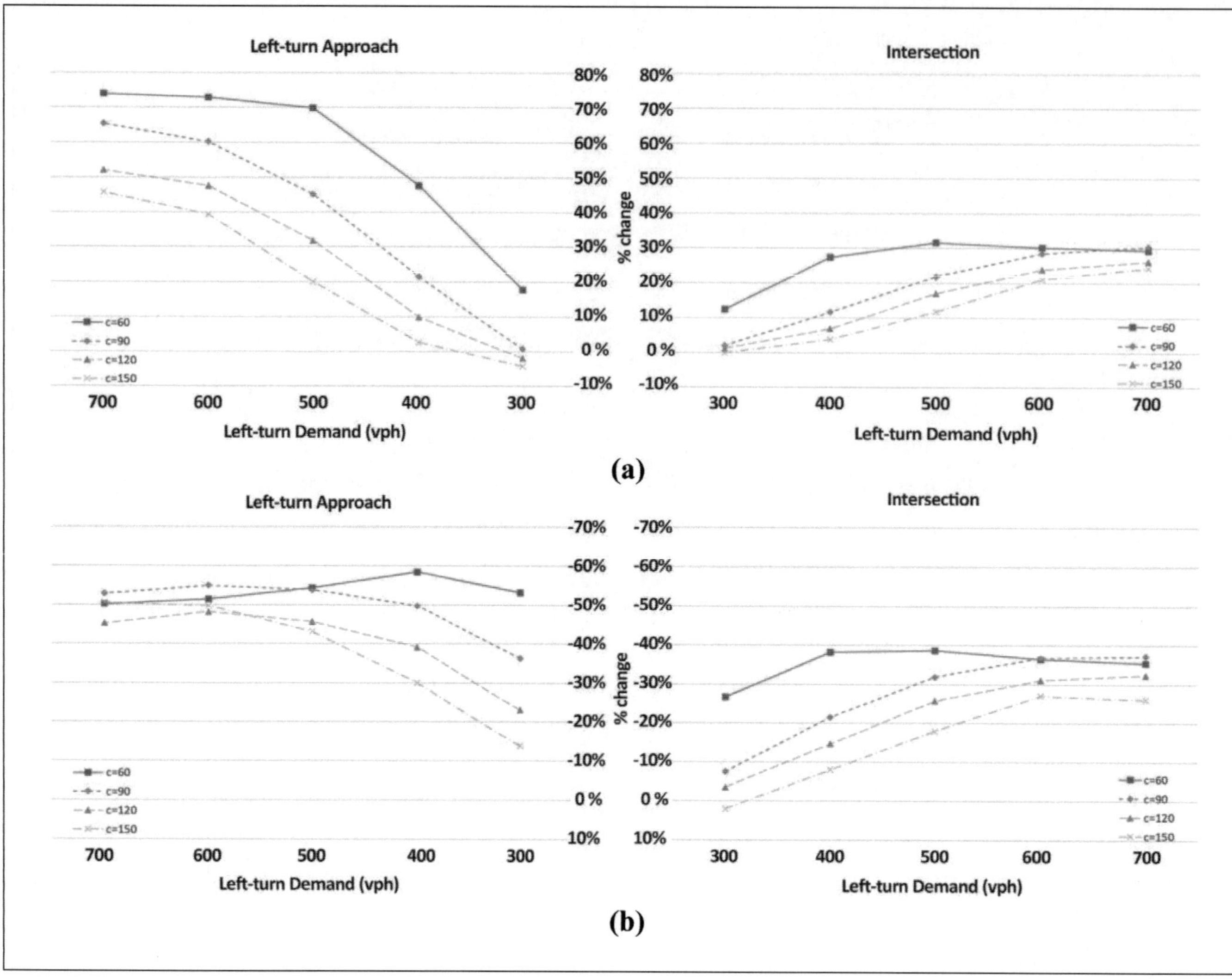

Figure 4. The impact of cycle length on MOEs: (*a*) % change of throughput for LT approach and intersection and (*b*) % change of delay for LT approach and intersection.

CLPL design reaches its capacity, vehicle delay reduction and throughput increases only change marginally.

Capacity Estimation Model

In this section, the capacity estimation model developed in the previous section is compared with the results obtained through simulation. The throughput obtained from the proposed model is slightly lower than estimations obtained from simulation alone, illustrating the importance of the proposed model. Additionally, the root mean squared error (RMSE) is less than 10% in all congested scenarios (see Figure 6), which validates the effectiveness of the proposed model in predicting the capacity of the CLPL implementation.

The advantage of this proposed model over pure simulation is that one may specify a probability that the CLPL will be used properly when it is open. Table 3 shows the suggested intersection capacity increase for a single exclusive LT approach with different cycle lengths (90 s, 120 s, 150 s) for different sets of the LT demand levels, and varying selection ratios of the CLPL (0.4, 0.45, 0.5). The following data can be used for future CLPL design plans.

Practical Implementation of CLPL

There are at least 50 real-world implementations of CLPL intersections in nine Chinese cities [for photos of implementations in China, see (*9*)]. Although no safety data were available, all three cities report positive impacts to the operations of the configured intersections. In Handan, throughput is estimated to increase to 720 vph. In Shenzhen, capacity increased by 30–40% for LT

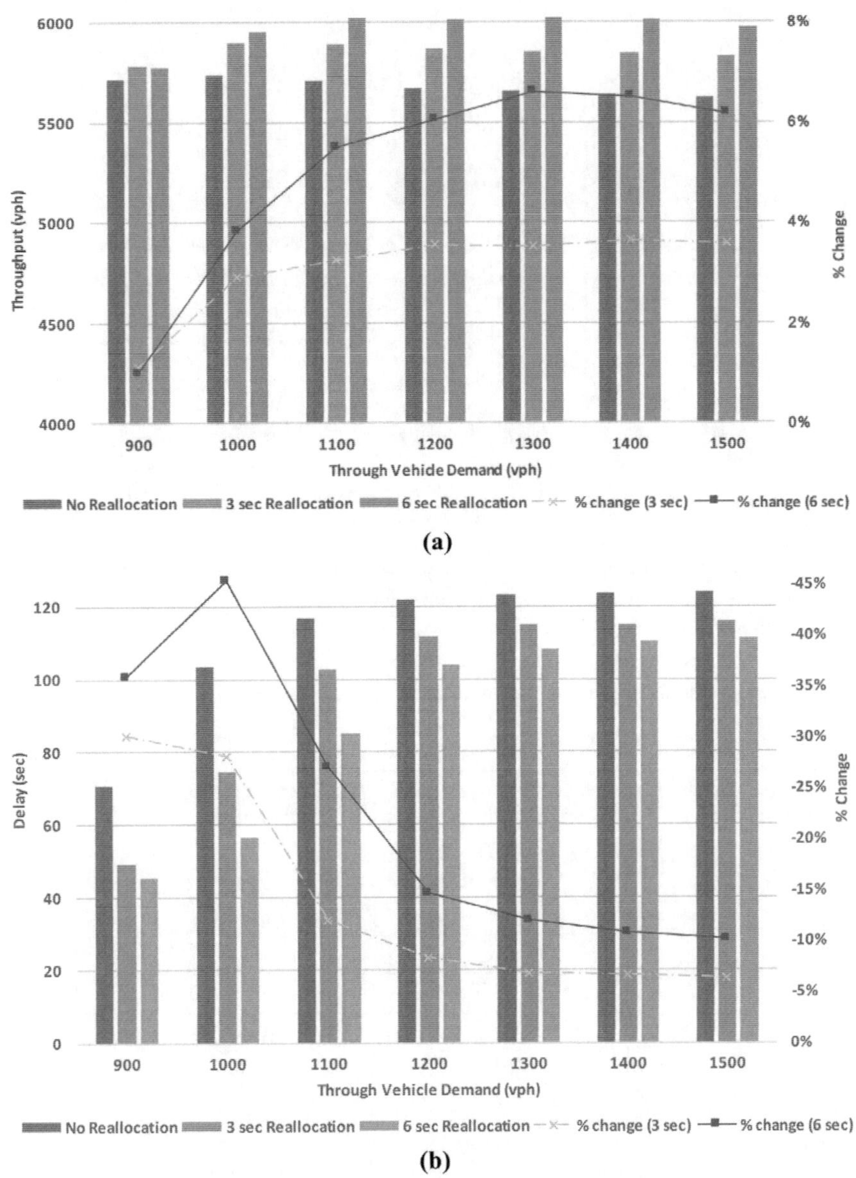

Figure 5. Change in MOEs as a function of time shifted from protected LTs to through vehicles: (*a*) throughput changes and (*b*) delay changes.

movements and 6% for the intersection. In Hohhot, capacity increased by 80% for LT movements (*9*).

FHWA's Office of Research and Development is conducting a human factors study at TFHRC for an existing intersection in Rockville, Maryland. This study is charged with developing appropriate signage for CLPL implementation, as well as determining the likelihood that the CLPL will be used appropriately (e.g., use CLPL when it is advantageous to do so, use the correct receiving lane, etc.) by a group of drivers experiencing the intersection design for the first time. The preliminary signage study has been completed and will soon be implemented in the human driving simulator (HDS). Figure 7*b* and *c* show the current iteration of proposed signage for the peak (open) and peak (closed)/off-peak conditions; an aerial view of the intersection in the simulation is illustrated in Figure 7*a*. The recommended peak signage when the CLPL is open is the combination of a positive contrast W12-1 sign on the median nose with appropriate lane use control signals (see Figure 7*b*). The recommended off-peak signage at the median nose is a positive contrast R4-7b sign (see Figure 7*c*); the most well understood set of signs will be tested in the HDS at TFHRC. A technical report on the results of this study

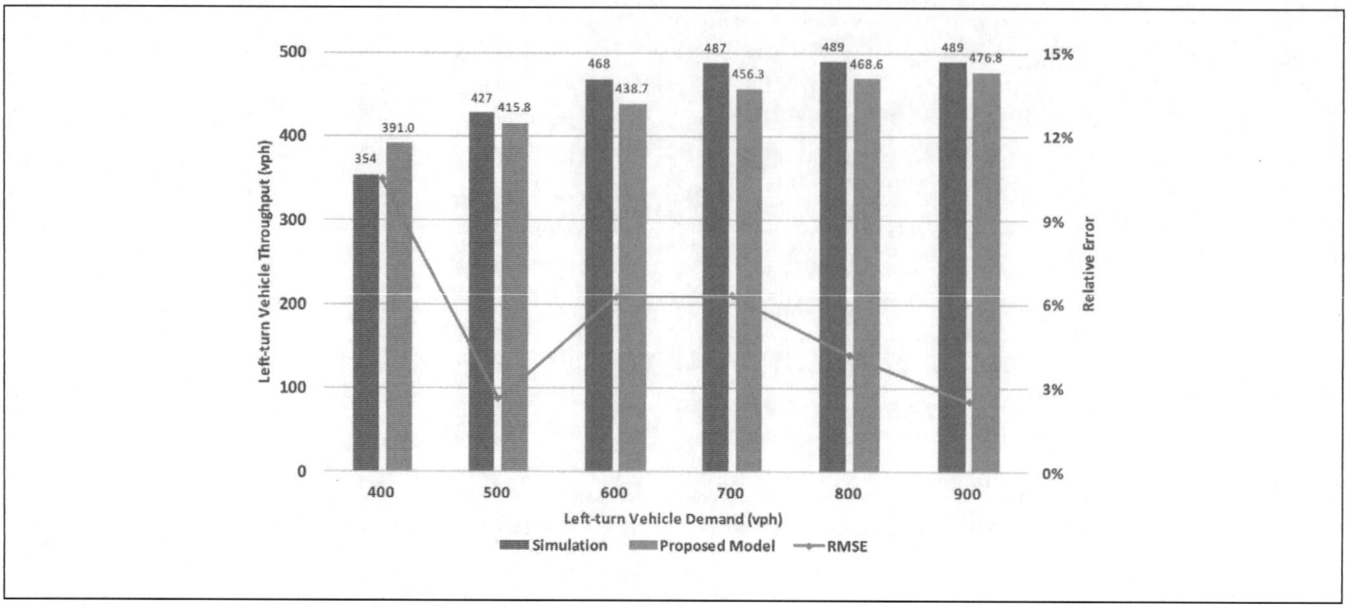

Figure 6. Estimation error between analytical model and simulation model.
*If queue length for CLPL is shorter than RLPL, then $p = 0.1$, otherwise, $p = 0.5$.

Table 3. Potential Capacity Increase under Different Demand and Cycle Combinations

C = 90 s			
LT demand	CLPL ratio $p = 0.4$	CLPL ratio $p = 0.45$	CLPL ratio $p = 0.5$
1–1.3 v/c ratio	42.2%	43.9%	45.1%
1.3–1.5 v/c ratio	51.2%	52.5%	53.3%
1.5–1.8 v/c ratio	59.5%	60.5%	61.1%
<1.8 v/c ratio	65.9%	66.6%	67.0%
C = 120 s			
LT demand	CLPL ratio $p = 0.4$	CLPL ratio $p = 0.45$	CLPL ratio $p = 0.5$
1–1.3 v/c ratio	36.5%	39.5%	40.6%
1.3–1.5 v/c ratio	48.5%	49.6%	50.4%
1.5–1.8 v/c ratio	58.8%	59.7%	60.3%
<1.8 v/c ratio	67.3%	67.9%	68.3%
C = 150 s			
LT demand	CLPL ratio $p = 0.4$	CLPL ratio $p = 0.45$	CLPL ratio $p = 0.5$
1–1.3 v/c ratio	23.2%	24.3%	25.0%
1.3–1.5 v/c ratio	27.1%	28.0%	28.5%
1.5–1.8 v/c ratio	35.2%	35.6%	35.8%
<1.8 v/c ratio	50.7%	50.9%	51.0%

is expected in 2018 from the Office of Safety Research and Development.

Conclusions and Recommendations

In conclusion, this study develops an analytical probability-based model for estimating the capacity of the CLPL design. The model considers many characteristics of a candidate intersection including signal timing, LT volume, vehicle arrival sequence, driver lane preference ratio, and the length of the CLPL. A comprehensive simulation study is conducted to test the benefit of the CLPL under various conditions. Additionally, the effectiveness of the developed analytical capacity estimation model is explored.

In general, the implementation benefits for CLPL can be summarized as:

- LT capacity increased by up to 70%.
- Delay reduction by up to 50% for LT traffic and 35% for all traffic using the intersection.

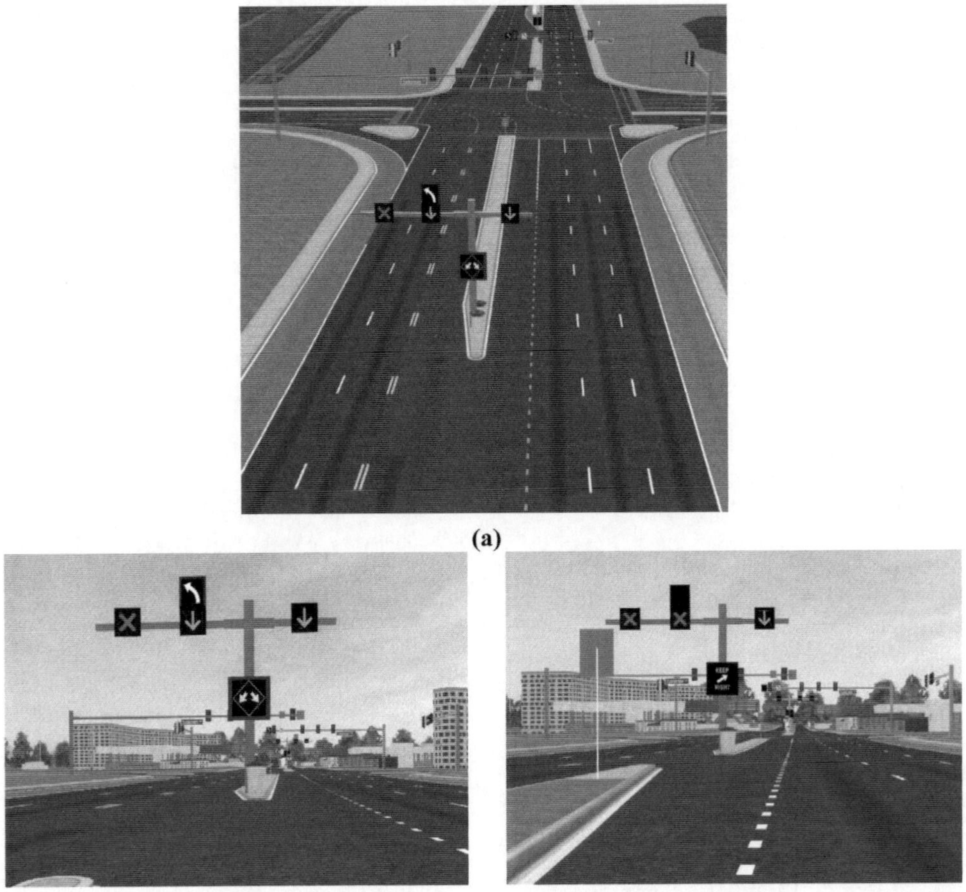

(a) (b) (c)

Figure 7. Proposed signage for CLPL implementation in traffic simulator: (*a*) aerial view of simulated intersection, (*b*) CLPL on-peak signage (CLPL open) (positive contrast W12-1 on median nose), and (*c*) CLPL on-peak signage (CLPL closed) (positive contrast R4-7b on median nose).

- The CLPL design can shift superfluous LT green time to heavy through traffic; 5% delay reduction can be expected with every 3 s shifted when v/c ratio is 1.

The available guidance regarding CLPL recommendation can be summarized as:

- The recommended cycle length ranges from 90 to 120 s.
- The recommended storage length ranges from 200 to 300 ft (8–12 vehicles).
- The CLPL can only run with leading protected-only LT phase sequences for both major and cross-streets.
- Cross-streets need to have at least two receiving lanes (i.e., individual receiving lane for CLPL and RLPL).
- Drivers will likely need time to adjust to the design as it breaks many traffic norms.
- CLPL may create potential traffic conflicts if not signed appropriately.
- Currently, a positive contrast W12-1 (on-peak, open) and R4-7b (on-peak, closed; off-peak) lane control signs on the median opening nose with appropriate overhead lane use control signals are being tested for comprehension at TFHRC. Recommendations from this human factors study on alternative lane groupings will be available in 2018.

Acknowledgments

This research was supported by the Pathways Program internship and was performed at the Turner-Fairbank Highway Research Center in McLean, VA. The authors also want to thank Dr. Yubian Wang from HRDS office at FHWA, who provided very insightful comments on an early version of the manuscript.

Author Contributions

The authors confirm contributing to the paper as follows: study conception and design: Yi Zhao and Joe Bared; data collection: Yi Zhao, Rachel James, and Lin Xiao; analysis and interpretation of results: Rachel James and Yi Zhao; draft manuscript preparation: Rachel James and Yi Zhao. All authors reviewed the results and approved the final version of the manuscript.

References

1. Jolovic, D., P. T. Martin, and A. Stevanovic. Left-turn Phasing—State-of-the-art Review and the Outlook for Future Improvements. *Proc., 16th COTA International Conference of Transportation Professionals*, 2016, pp. 1624–1644, http://ascelibrary.org/doi/10.1061/9780784479896.149
2. Choi, E. H. *Crash Factors in Intersection-related Crashes: An On-scene Perspective*. NHTSA Technical Report, 2010, http://www.distraction.gov/downloads/pdfs/crash-factors-in-intersection-related-crashes.pdf
3. Lord, D., A. Smiley, and A. Haroun. Pedestrian Accidents with Left-turning Traffic at Signalized Intersection: Characteristics, Human Factors, and Unconsidered Issues. Presented at 77th Annual Meeting of the Transportation Research Board, Washington, D.C., 1998.
4. Rodegerdts, L., K. Lee, S. Quayle, S. Beaird, C. Braud, J. Bonneson, P. Tarnoff, and T. Urbanik. *Traffic Signal Timing Manual*. Technical Report: FHWA-HOP-08-024. 2008.
5. *Highway Capacity Manual 2010*. Transportation Research Board of the National Academies, Washington, D.C., 2010.
6. Haddad, J., and N. Geroliminis. Effect of Left Turns for Arterials with Queue Spillbacks. Presented at 92nd Annual Meeting of the Transportation Research Board, Washington, D.C., 2013.
7. Skabardonis, A., and N. Geroliminis. Real-time Monitoring and Control on Signalized Arterials. *Journal of Intelligent Transportation Systems*, Vol. 12, No. 2, 2008, pp. 64–74.
8. Ren, Y., Y. Wang, G. Yu, H. Liu, and L. Xiao. An Adaptive Signal Control Scheme to Prevent Intersection Traffic Blockage. *IEEE Transactions on Intelligent Transportation Systems*, Vol. PP, No. 99, 2016, pp. 1–10.
9. Hale, D., R. Jagannathan, M. Xyntarakris, P. Su, X. Jiang, J. Ma, J. Hu, and C. Krause. *Traffic Bottlenecks: Identifications and Solutions*. FHWA Technical Report. 2016.
10. Zhao, J., M. Yun, H. M. Zhang, and X. Yang. Driving Simulator Evaluation of Drivers' Response to Intersections with Dynamic Use of Exit-lanes for Left-turns. *Accident Analysis and Prevention*, Vol. 81, 2015, pp. 107–119.
11. Wainwright, S. Traffic Safety Toolbox: A Primer on Traffic Safety One-Way Streets and Reversible Lanes. *Institute of Transportation Engineers*, Vol. 141, 1993, 147p.
12. Wolshon, B., and L. Lambert. Reversible Lane Systems: Synthesis of Practice. *Journal of Transportation Engineering*, Vol. 132, No. 12, 2006, pp. 933–944.
13. Institute of Transportation Engineers; Wolshon, B., and A. Pande. *Traffic Engineering Handbook*, 7th ed. John Wiley & Sons, Inc., 2016.
14. Wolshon, B., and L. Lambert. *Convertible Roadways and Lanes. National Cooperative Highway Research Program*, Synthesis 340, 2004. http://nap.edu/23331
15. Cao, Y., Z. Zuo, and H. Xu. Analysis of Traffic Conflict Characteristic at Temporary Reversible Lane. *Periodica Polytechnica Transportation Engineering*, Vol. 42, No. 1, 2013, pp. 73–76.
16. Lambert, L., and B. Wolshon. Characterization and Comparison of Traffic Flow on Reversible Roadways. *Journal of Advanced Transportation*, Vol. 44, 2010, pp. 113–122.
17. Zhao, J., W. Ma, M. Zhang, and X. Yang. Increasing the Capacity of Signalized Intersections by Dynamically Using Exit-lanes for Left-turn Traffic. *Transportation Research Record: Journal of the Transportation Research Board*, 2013. 2355: 49–59.
18. Krause, C., D. Hale, J. Bared, and Z. Huang. Operational Advantages of Contraflow Left-turn Pockets at Signalized Intersections. *Institute of Transportation Engineers. ITE Journal*, Vol. 86, No. 7, 2016, 44p.
19. Wu, J., P. Liu, Z. Tian, and C. Xu. Operational Analysis of the Contraflow Left-turn Lane Design at Signalized Intersections in China. *Transportation Research Part C*, Vol. 69, 2016, pp. 228–241.
20. Tian, Z. Z., and >N. Wu. Probabilistic Model for Signalized Intersection Capacity with a Short Right-turn Lane. *Journal of Transportation Engineering*, Vol. 132, No. 3, 2006, pp. 205–212.
21. Wu, N. Total Approach Capacity at Signalized Intersections with Shared and Short Lanes: Generalized Model Based on a Simulation Study. *Transportation Research Record: Journal of the Transportation Research Board*, 2007. 2027: 19–26.
22. Kikuchi, S., and N. Kronprasert. Determining the Length of the Right-turn Lane at a Signalized Intersection. *Transportation Research Record: Journal of the Transportation Research Board*, 2008. 2060: 19–28.
23. Kikuchi, S., and N. Kronprasert. Determining Lengths of Left-turn Lanes at Signalized Intersections Under Different Left-turn Signal Schemes. *Transportation Research Record: Journal of the Transportation Research Board*, 2010. 2195: 70–81.

The Standing Committee on Operational Effects of Geometrics (AHB65) peer-reviewed this paper (18-05615) and selected it to receive its 2018 Best Paper Award.

Article

TRR
JOURNAL OF THE TRANSPORTATION RESEARCH BOARD

Transportation Research Record
2018, Vol. 2672(17) 35–47

DOI: 10.1177/0361198118783152
journals.sagepub.com/home/trr
SAGE

Modeling the Risk of Wrong-Way Driving Entry at the Exit Ramp Terminals of Full Diamond Interchanges

Md Atiquzzaman[1] and Huaguo Zhou[1]

Abstract

Wrong-way driving (WWD) crashes are a critical safety issue on freeways. Although these crashes are rare and random in nature, they often result in severe injuries and/or fatalities. Typically, exit ramp terminals are the initial point of entry for wrong-way (WW) drivers on freeways. Therefore, it is important for transportation agencies to identify the exit ramp terminals with higher possibility of WW entries and apply safety countermeasures to reduce the chances of their occurrence. However, the random nature of WWD crashes and the difficulty in identifying the actual entry points makes it hard for transportation agencies to assess the risk of WWD at a particular exit ramp terminal and apply countermeasures accordingly. This study developed mathematical models for predicting the risk of WW entries at the exit ramp terminals of full diamond interchanges. The geometric design features, usage of traffic control devices, area type where the interchanges are located, and annual average daily traffic (AADT) at the exit ramp terminals with or without history of WWD were used as potential predictors of WW entry. Transportation agencies can use these models to assess the risk of WW entries at the exit ramp terminals within their jurisdictions and consider possible countermeasures. They also can be applied during the design phase to determine the combination of geometric design features and traffic control devices that ensures the least possibility of WW entry.

Wrong-way driving (WWD) is the act of driving in the opposite direction to legal traffic flow on relatively high-speed divided highways or access ramps (*1*). Although the frequency of WWD-related collisions is rare compared with other types of highway crashes, they are typically associated with the higher possibility of severe injuries and fatalities. During the period of 2004–2011, an average of 269 WWD fatal crashes resulted in 359 fatalities annually in the United States, which accounts for 1.33 fatalities per WWD fatal crash compared with 1.10 fatalities for all types of fatal crashes (*2*). This is because the majority of WWD crashes are head-on events. Thus, WWD crashes have drawn a lot of attention from transportation professionals over the past decades despite the fact that they are relatively infrequent events on our roadways.

Typically, entering controlled access highways through an exit ramp is the primary origin of WWD movements. Although WWD movements sometimes originate from making a U-turn on the mainline and at median crossover, they account for only a small portion of all WWD events. The mechanism of how a driver can make a U-turn on freeways to end up driving the wrong way is explained in (*1*, *3*). A study in Illinois showed that only 6.5% of 217 confirmed WWD crashes occurred from wrong-way (WW) drivers making a U-turn on the mainline, while the other 93.5% likely occurred from drivers entering the freeway through an exit ramp (*3*). Therefore, the geometric design characteristics and usage of traffic control devices (TCDs) on the exit ramp terminals as well as along the ramps are critical in reducing the occurrence of WWD crashes on freeways. This study focuses on developing mathematical models to predict the probability of WW entry at the exit ramp terminals of full diamond interchanges, primarily based on geometric design features and usage of TCDs.

Diamond interchanges are the most widely used service interchange (79% of all interchanges fall in this category) in the United States (*4*). Although they are less susceptible to WWD events compared with partial cloverleaf or trumpet interchanges (*5*–*8*), the origin of a

[1]Department of Civil Engineering, Auburn University, Auburn, AL

Corresponding Author:
Address correspondence to Md Atiquzzaman: mza0113@auburn.edu

major portion of WWD crashes can be attributed to the exit ramp terminals of diamond interchanges (*3*). This is due to the high number of diamond interchange presence in the United States' freeway facilities. Therefore, the assessment of WWD risk at the exit ramp terminals of full diamond interchanges using mathematical models will greatly benefit the state and local transportation agencies in identifying the high-risk exit ramp terminals and applying safety countermeasures to deter WWD.

The remainder of this article is organized as follows: (1) a summary of relevant WWD studies; (2) the data collection process; (3) the methodology used to develop the mathematical models; (4) the results obtained by modeling the probability of WWD entry at the exit ramp terminals of full diamond interchanges; and (5) key findings on geometric features and TCDs that can affect the risk of WW entries onto off-ramps of full diamond interchanges.

Literature Review

Identifying the contributing factors in WWD crashes has been a primary focus of much WWD research for the past several decades. Numerous studies revealed some common factors that contribute to the occurrence of WWD crashes. These studies collectively concluded that older drivers, younger drivers, male drivers, driving under the influence of alcohol or drugs, poor lighting conditions, urban areas, early morning hours, weekend days, and severe weather conditions contribute to WWD crashes (*3*, *6*, *8–13*). Additionally, a few of these studies identified that some interchange types are more susceptible to cause driver confusion and contribute to WW entries. For instance, Copelan reported that the interchanges with short sight distance, partial cloverleaf interchanges, trumpet interchanges, half and full diamond interchanges, buttonhook ramps, slip ramps, four-legged intersection near exit ramps, left-side exit ramps, and scissors exit ramps are more likely to cause WW entries (*9*). Cooner et al. reported left-side exit ramps and one-way streets transitioning into a freeway are likely to cause more WWD (*10*). Additionally, previous studies reported the two-quadrant partial cloverleaf, trumpet, tight diamond, and full diamond interchanges to be more susceptible to WWD (*6*, *8*).

Only a handful of previous studies attempted to predict WWD crash risk at certain interchanges, road segments, or within specific jurisdiction. For example, Baratian-Ghroghi et al. developed a mathematical model to predict the probability of WWD incidents at the exit ramp terminals of partial cloverleaf interchanges, in which they used six-hours of traffic movement at a signalized exit ramp terminal and estimated the number of potential WWD maneuvers using VISSIM simulation software (*14*). Their findings show that the WWD events increase when left-turn volume toward entrance ramp increases and stopping vehicles at exit ramp decreases. Sandt et al. identified WWD hotspots by modeling crash risk and analyzing traffic management response times (*15*). They developed a Poisson regression model to predict WWD crashes at different roadway segments in the Central Florida area based on WWD event data (e.g., crashes, citations, and 911 calls with GPS location) and roadway data (e.g., interchange locations, interchange types, and traffic volumes). Additionally, they identified WWD hotspots based on the time spent responding to WWD events, which can be used if accurate WWD 911 calls and citation data with GPS location are not available. Similar WWD crash risk models were developed for the South Florida area (*16*). Earlier, Rogers et al. conducted a study to model the risk of WWD crashes for interstates/toll facilities and counties in Florida based on statewide WWD crashes, citations, and 911 calls (*17*).

Although a few studies attempted to predict the risk of WWD crashes, the focus was mainly on macroscopic level. To the best of our knowledge, only one study explored the contribution of microscopic level geometric design features on WW entries (*18*). However, this study was conducted for only the exit ramp terminals of partial cloverleaf interchanges, and the sole aim was to identify the geometric design features that contribute to WW entries instead of modeling the risk of WW entries. Therefore, none of the previous studies attempted to predict the risk of WWD crashes at the exit ramp terminals of full diamond interchanges, which is the primary origin of a large portion of WWD events. The scarcity of research in this area can be attributed to the rareness and random nature of WWD events along with the difficulty in determining the true entry points of the WW drivers. To fill this gap, this study used the geometric design features, usage of TCDs, annual average daily traffic (AADT) data, and area type to predict the risk of WW entries at the exit ramp terminals of full diamond interchanges.

Data Collection

Identifying the exit ramp terminals with history of WWD entry is critical to model the risk of WW entries. In this study, the WWD crashes on Alabama and Illinois freeways were collected for a period of five years (2009–2013). Additionally, the interchanges near crash locations were located using Google maps. After that, the entry points of all WWD crashes during the study period were determined through a careful review of crash narratives and the methodology proposed by Zhou et al. (*3*). Altogether, a total of 128 exit ramp terminals of full diamond interchanges were identified to have at least one

Table 1. Descriptive Summary of Geometric Design Features and TCD Application Practices at Exit Ramp Terminals of Full Diamond Interchanges

	Locations with history of WWD (n = 128, 23%)		Locations without history of WWD (n = 428, 77%)	
Variable	Frequency	Percent	Frequency	Percent
Intersection angle				
Acute	35	27.34	54	12.62
Right	41	32.03	125	29.21
Obtuse	52	40.63	249	58.18
Median on crossroad				
Non-traversable	55	42.97	277	64.72
Traversable	73	57.03	151	35.28
Channelizing island				
Non-traversable	72	56.25	307	71.73
Traversable	56	43.75	121	28.27
Distance to access point				
200 ft and less	13	10.16	38	8.88
201–400 ft	36	28.13	84	19.63
401–600 ft	29	22.66	67	15.65
601–800 ft	15	11.72	119	27.80
More than 800 ft	35	27.34	120	28.04
Is corner radius tangent to crossroad?				
Yes	69	53.91	127	29.67
No	59	46.09	301	70.33
Number of lanes on the crossroad				
2 or 3 lanes	55	42.97	270	63.08
4 or more lanes	73	57.03	158	36.92
Number of lanes on exit ramp				
One	29	22.66	53	12.38
Two	68	53.13	313	73.13
Three or more	31	24.22	62	14.49
Distance of WW-sign from crossroad				
200 ft and less	32	25.00	280	65.42
201–300 ft	40	31.25	70	16.36
301–400 ft	19	14.84	24	5.61
401–500 ft	18	14.06	28	6.54
More than 500 ft	19	14.84	26	6.07
Usage of DNE sign				
One (right/left side of exit ramp)	10	7.81	29	6.78
Two (channelizing island and right/left side of exit ramp)	26	20.31	61	14.25
Two (both side of exit ramp)	78	60.94	319	74.53
Three (channelizing island and both side of exit ramp)	14	10.94	20	4.67
Signalized/unsignalized?				
Signalized	54	42.19	127	29.67
Unsignalized	74	57.81	301	70.33
Area type				
Rural	49	38.28	268	62.62
Urban	79	61.72	160	37.38

WW entry during the study period. These exit ramp terminals were further investigated using Google Earth aerial and street views to collect the geometric design features and TCD usage information. Initially, the 2017 Google Earth imagery was used for data collection. Then they were cross-checked with the imagery of crash year or closest year available to verify if there were any improvements made at the study locations. In case there were any improvements, the data was adjusted to ensure that it portrays the crash year's geometric and signage conditions. As a comparison group, similar information was collected for 428 exit ramp terminals with no history of WWD during the study period. A descriptive summary of collected data is presented in Table 1.

The collected geometric design features include intersection angle, type of median on crossroad, type of channelizing island, distance of nearest access point from the exit ramp terminals, tangency of corner radius to the crossroad edge, and number of lanes on the exit ramps and crossroads. Based on the past studies and existing geometric guidelines, these geometric design features have considerable effect on the possibility of WW entry (*19*, *20*).

Intersection angle is the angle between the centerline of a ramp and the centerline of crossroad median, measured from the right side of the ramp (*21*). In this study, the intersection angle was categorized as acute, right, and obtuse angle. Because previous studies revealed that a 5-degree deviation from right angle is typically indistinguishable by the drivers (*22*), we defined the intersection angle as follows: acute—less than 85 degrees; right—85 to 95 degrees; and obtuse—more than 95 degrees.

The type of median on crossroads and channelizing island on exit ramp are two important design features to reduce the probability of WW entries. A non-traversable raised median on crossroad makes the WW left-turn from a crossroad less likely (*20*). Similarly, a non-traversable channelizing island reduces the chance of WW entries by narrowing the exit ramp throat (*19*).

The presence of access points close to exit ramp terminals increases the chance of WW entries at partial cloverleaf interchanges (*18*). Thus, the distance of nearest access points from the exit ramp terminals was collected with an aim to understand their effect on WW entries at full diamond interchanges.

Existing guidelines stressed on using an angular connection at the intersection of the left edge of exit ramps and right edge of crossroads, thus making the corner radius non-tangent to crossroad edge, to discourage WWD (*20*, *23*). However, none of the existing literature quantitatively measured the effect of this geometric feature on WWD.

In addition to the geometric features discussed above, the number of lanes on an exit ramp and crossroad were collected as potential variables to be included in the model to predict the risk of WW entries. For a better understanding, Figure 1 provides a visual representation of the discussed geometric design features.

WW-related signs and intersection signalization at the exit ramp terminals are also critical for reducing WWD. However, the impact of WW signs placement, number of "Do Not Enter" (DNE) signs, and signalization on WWD is not clear. Hence, the research team collected the distance of WW signs from crossroads, the number and location of DNE signs at the exit ramp, and intersection signalization information to be included in the model to predict the risk of WW entry.

Furthermore, the area types (i.e., urban/rural) of the study interchanges were recorded using the 2010 Census Urban Area map, as the previous studies indicated that the interchanges in urban areas are typically over-representative in WWD crashes. Also, the AADT on the exit ramp and the crossroad may play a significant role in predicting the risk of WW entry. Therefore, AADT on the exit ramps and the crossroads were collected from Alabama Department of Transportation (ALDOT) and Illinois Department of Transportation (IDOT) traffic count websites.

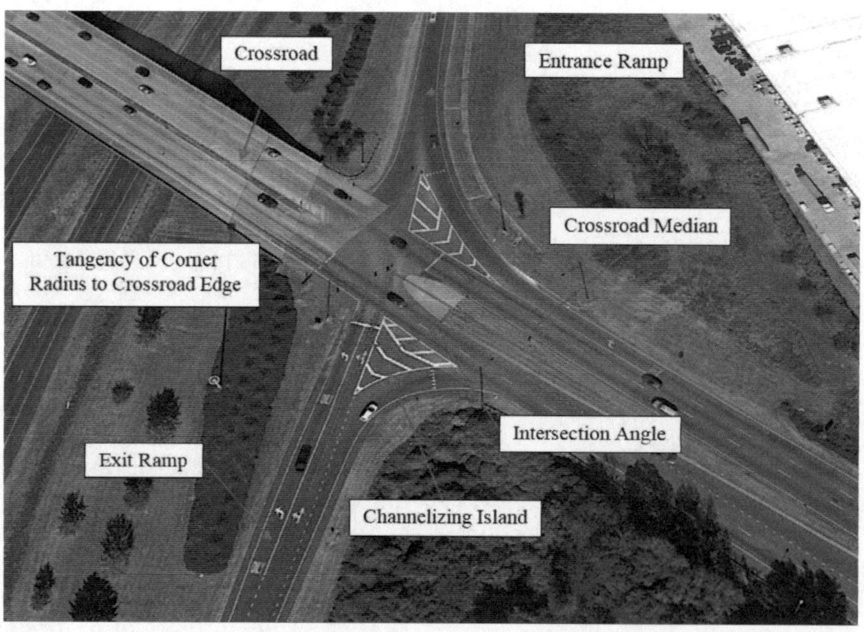

Figure 1. Geometric design features of full diamond interchange exit ramp terminals.

Methodology

The final dataset consisted of 128 exit ramp terminals with history of WW entries and 428 exit ramp terminals with no history of WW entries. Primarily, a model was developed using all 556 observations, which included the exit ramp terminals connected to both two-lane and multilane crossroads. Further, the final dataset was divided into two categories based on the number of lanes on the crossroads (i.e., two-lane and multilane). After that, separate models were developed for exit ramp terminals connected to two-lane and multilane crossroads. This was done to investigate if a particular geometric feature or TCD is more important for the exit ramp terminal connected to a two-lane crossroad than a multilane crossroad. Therefore, a total of three models were developed: Model 1—a model to predict WW entry regardless of the number of lanes on the crossroad, Model 2—a model to predict WW entry at the exit ramp terminal connected to a two-lane crossroad, and Model 3—a model to predict WW entry at the exit ramp terminal connected to a multilane (four or more lanes) crossroad.

The response variable in this study is dichotomous in nature (i.e., exit ramp terminals with or without a history of WW entries). Binary logistic regression is a classic statistical technique to handle such problems with dichotomous dependent variables. Moreover, many previous studies used this technique to study the probability of traffic crashes and the effects of different variables on certain types of crashes (*24–29*). While standard binary logistic regression, which is based on maximum likelihood method, works well for a large and balanced sample size, it may produce biased outcome and convergent failures when applied to rare event crash data (*30–33*). Therefore, standard binary logistic regression models may not be suitable for analyzing WWD events due to their rareness. To overcome these limitations, this study used Firth's penalized likelihood logistic regression method, which minimizes the biased probability and convergent failures resulting from the maximum likelihood estimate of rare event data.

To develop the model, the response variables were assigned a binary indicator that had a value of 0 if there were no history of WW entry at an exit ramp terminal and 1 otherwise. Since most of the locations in this study have only one WW event, binary response variable is a logical approach in this case. In the future, researchers can collect WW incidents (citations and 911 calls) to include number of incidents by locations to the model and develop a model to predict expected number of incidents instead of the probability of WW entry. The categorical explanatory variables were also assigned binary indicator of 0 or 1. Among the explanatory variables, AADT was the only continuous variable. The actual value of AADT is very large compared with the binary indicator (0 or 1), which can skew the model. Therefore, the AADT values were transformed to a logarithmic scale to reduce the skewness. The fitted logistic regression model can be expressed as shown in Equation 1:

$$logit(p) = \ln\left(\frac{p}{1-p}\right) = \beta_0 + \beta_1 X_1 + \beta_2 X_2 + \beta_3 X_3 + \ldots + \beta_n X_n \quad (1)$$

where p is the probability of WW entry at an exit ramp terminal, X_1, X_2, X_3.........X_n are the explanatory variables, β_0 is the intercept, and $\beta_1, \beta_2, \beta_3$, ...$\beta_n$ are the regression coefficients.

After preparing the dataset, a full logistic model was developed, which included all the explanatory variables in the primary dataset. However, not all the variables in primary dataset had a significant effect on predicting the probability of WW entry. Therefore, in the next step, the backward elimination technique was employed to achieve the most parsimonious logistic model and to find out the subset of variables that are most suitable in predicting the probability of WW entry. The backward elimination technique uses the Akaike information criteria (AIC), which can be calculated using Equation 2, to decide the most parsimonious model. Typically, the most parsimonious and best-fitted model is the one that produces lowest AIC.

$$\text{AIC} = 2k - L_\beta \quad (2)$$

where k is the number of explanatory variables, and L_β is the maximum log-likelihood of the model.

Finally, the odds ratio (OR) was computed for the explanatory variables included in the final models. By definition, OR of a certain variable expresses the change in the probability of WW entry caused by a unit change of that same variable, while other variables remain constant. The OR can range from 0 to infinity, where a value of greater than 1 indicates the increased probability of WW entry, and a value of less than 1 indicates the decreased probability.

Results and Discussion

Model 1: Base Model for Predicting WW Entry

As previously mentioned, all the observations in the final dataset (i.e., 128 exit ramps with history of WW entries and 428 with no history of WW entry) were included in the base model. The aim was to develop a model that can predict the probability of WW entry regardless of the number of lane(s) on connecting crossroad. First, a full model was developed considering all the explanatory

Table 2. Summary of Firth's Logistic Regression for All Ramp Terminals (Model 1)

Variables	Parameter estimates	Estimated standard error	Chi-square statistics	OR
Intercept	−4.799	1.381	13.144	
Intersection angle				
Right	Reference			
Acute	0.257	0.320	0.659	1.293
Obtuse	−0.644	0.268	5.823	0.525
Median on crossroad				
Non-traversable	Reference			
Traversable	0.233	0.304	0.597	1.262
Is corner radius tangent to crossroad edge?				
No	Reference			
Yes	1.309	0.303	20.373	3.702
Distance of WW sign from crossroad				
200 ft. and less	Reference			
201–300 ft.	1.313	0.307	18.600	3.717
301–400 ft.	1.634	0.388	17.592	5.124
401–500 ft.	1.448	0.394	13.344	4.255
More than 500 ft.	1.576	0.403	15.325	4.836
Signalized/unsignalized?				
Unsignalized	Reference			
Signalized	−0.305	0.372	0.687	0.737
log(exit ramp AADT)	−0.410	0.400	1.066	0.664
log(crossroad AADT)	0.832	0.435	3.883	2.298
Area type				
Rural	Reference			
Urban	1.361	0.357	15.022	3.900

variables. However, not all the explanatory variables were significantly capable of predicting the probability of WW entry. Therefore, a backward elimination technique was employed to achieve a reduced final model, which only considers a subset of explanatory variables in the preliminary dataset and minimizes the AIC value. A Chi-square test between the null model and the final model indicated that the deviance of the final model is significantly lower ($p < .001$) than the null model.

The summary of final base model is presented in Table 2. The fitted logistic regression model is shown in Equation 3. In the fitted model, p is the probability of WW entry at exit ramp terminals of a full diamond interchange.

$$\textbf{logit}(p) = \textbf{ln}\left(\frac{p}{1-p}\right) = -4.799 + 0.257(X_{\text{IA1}}) - 0.644(X_{\text{IA2}}) + 0.233(X_{\text{MC}}) + 1.309(X_{\text{CR}}) + 1.313(X_{\text{WWSD1}}) + 1.634(X_{\text{WWSD2}}) + 1.448(X_{\text{WWSD3}}) + 1.576(X_{\text{WWSD4}}) - 0.305(X_{\text{Signal}}) - 0.410(X_{\text{log(exit ramp AADT)}}) + 0.832(X_{\text{log(crossroad AADT)}}) + 1.361(X_{\text{AT}}) \tag{3}$$

Here,

$$X_{\text{IA1}} = \begin{cases} 1, & \text{if intersection angle is acute} \\ 0, & \text{otherwise} \end{cases}$$

$$X_{\text{IA2}} = \begin{cases} 1, & \text{if intersection angle is obtuse} \\ 0, & \textit{otherwise} \end{cases}$$

$$X_{\text{MC}} = \begin{cases} 1, & \text{if median on crossroad is traversable} \\ 0, & \text{otherwise} \end{cases}$$

$$X_{\text{CR}} = \begin{cases} 1, & \text{if corner radius is tangent to the edge of crossroad} \\ 0, & \text{otherwise} \end{cases}$$

$$X_{\text{WWSD1}} = \begin{cases} 1, & \text{if the distance of WW sign from crossroad is between 201 ft to 300 ft} \\ 0, & \text{otherwise} \end{cases}$$

$$X_{\text{WWSD2}} = \begin{cases} 1, & \text{if the distance of WW sign from crossroad is between 301 ft to 400 ft} \\ 0, & \text{otherwise} \end{cases}$$

$$X_{\text{WWSD3}} = \begin{cases} 1, & \text{if the distance of WW sign from crossroad is between 401 ft to 500 ft} \\ 0, & \text{otherwise} \end{cases}$$

$$X_{\text{WWSD4}} = \begin{cases} 1, & \text{if the distance of WW sign from crossroad is more than 500 ft} \\ 0, & \text{otherwise} \end{cases}$$

$$X_{\text{Signal}} = \begin{cases} 1, & \text{if the exit ramp terminal is signalized} \\ 0, & \text{otherwise} \end{cases}$$

$X_{\text{log(exitrampAADT)}}$ = Logarithmic value of exit ramp AADT

$X_{\text{log(crossroad AADT)}}$ = Logarithmic value of crossroad AADT

$$X_{\text{AT}} = \begin{cases} 1, & \text{if the interchange is located in urban area} \\ 0, & \text{otherwise} \end{cases}$$

Equation 3 indicates that the intersection angle, median on crossroad, tangency of corner radius to crossroad edge, distance of WW sign from crossroad, intersection signalization, exit ramp AADT, crossroad AADT, and area type were significantly capable of predicting the probability of WW entry. Other explanatory variables (e.g., distance to nearest access point, channelizing island, and usage of DNE sign) were not capable of predicting WW entry significantly and thus excluded from the final model. In Equation 3, the negative signs before the independent variables such as obtuse angle connection, traffic signal, and exit ramp AADT indicate that they reduce the chances of WW entry, and the positive sign indicates that they increase the chances of WW entry. Additionally, the OR values, as reported in Table 2, explain the extent to which each parameter is responsible for increasing or decreasing the chances of WW entry.

According to the developed model, the design features such as the presence of obtuse intersection angle, signalized ramp terminal, and higher exit ramp AADT reduce the chances of WW entry. These are some interesting results. The AASHTO design guideline recommended using a right-angle connection between one-way exit ramps of full diamond interchanges and the crossroads. However, the results here showed that, for a full diamond interchange, obtuse intersection angle reduces the probability of WW entry compared with a right/acute-angle connection. The reason behind this may be attributed to the fact that WW right-turns are more prevalent at the exit ramp terminals of full diamond interchanges, while an obtuse-angle connection makes the WW right-turns very difficult. On the contrary, the acute angle makes the right-turning maneuver easy and therefore found to be more prone to WWD (OR = 1.293). For a better understanding, a visual representation of WW right-turning maneuver is shown in Figure 2.

The odds of WW entry slightly increased for a traversable median on the crossroad (OR = 1.262). This result clearly supports the *AASHTO Green Book*'s recommendations for using non-traversable crossroad medians to deter the WW left-turning movements from crossroads (*20*).

The *AASHTO Green Book* (*20*) recommended that the corner radius should not be tangent to the crossroad edge, thus making the connection between left edge of exit ramp and right edge of crossroad angular. Refer to Figure 3 for an example of corner radius tangent and non-tangent to the crossroad edge. The results show that the use of corner radius tangent to crossroad edge increases the odds of WW entry by 3.7 times, which can be attributed to the fact that angled corner makes the WW right-turning maneuvers difficult. Therefore, our result supports the *Green Book* guidance for using angled corner to deter right-turning WW entry from crossroads.

The Manual on Uniform Traffic Control Devices (MUTCD) (*34*) requires the use of at least one WW sign on the exit ramps. However, there is no specific guidance on proper placement of WW sign along the exit ramps. The results show that the odds of WW entry increase by three to five times when the distance between the first WW sign and crossroad is more than 200 ft compared to when the distance is 200 ft or less. This is an interesting result given that the placement of WW sign varies widely among the state and local transportation agencies and there is no available guideline on the proper location for placing the WW sign on the exit ramps. Based on the results, the favorable distance of first WW sign from crossroad is 200 ft or less. A more conservative approach may include the use of a second set of WW signs close to freeway and exit ramp connections for the drivers who somehow miss the first WW signs. It should be noted that some states already use two sets of WW signs along the exit ramp (the first set close to exit ramp terminal and a second set farther away).

Figure 2. WW right-turning maneuvers for (*a*) obtuse angle and (*b*) acute angle connection.

Figure 3. Examples of (*a*) corner radius tangent to crossroad edge and (*b*) corner radius not tangent to crossroad edge or corner radius tangent to crossroad median.

Additionally, the developed model shows that the signalized exit ramp terminals are safer in terms of WW entry compared with unsignalized intersections, which is understandable given the fact that the signalized intersections facilitate more controlled and regulated traffic movements. Based on this finding, the probability of WW entry is recommended to be considered as a supplement to the MUTCD traffic signal warrants to justify intersection signalization at the exit ramp terminals.

Finally, the crossroad AADT showed a positive impact on the probability of WW entry at exit ramp terminals of full diamond interchanges, which means that the higher crossroad AADT increases the chance of WW entry. This is no surprise, as the higher AADT means more potential WW drivers. Sandt et al. also found that the higher AADT value increases the risk of WWD crashes when predicting WWD crash risk on different road segments in Central Florida (*15*).

Table 3. Summary of Firth's Logistic Regression Model for Exit Ramp Terminals Connected to Two-Lane Crossroads (Model 2)

Variables	Parameter estimates	Estimated standard error	Chi-square statistics	OR
Intercept		−2.364	1.835	1.759
Intersection angle				
Right	Reference			
Acute	−0.264	0.507	0.284	0.768
Obtuse	−1.145	0.387	9.069	0.318
Is corner radius tangent to crossroad edge?				
No	Reference			
Yes	2.181	0.461	29.802	8.855
Distance of WW sign from crossroad				
200 ft. and less	Reference			
201–300 ft.	1.153	0.499	5.152	3.168
301–400 ft.	1.460	0.615	5.429	4.306
401–500 ft.	1.285	0.509	6.409	3.615
More than 500 ft.	1.374	0.509	7.328	3.951
Signalized/unsignalized?				
Unsignalized	Reference			
Signalized	−0.251	0.879	0.087	0.778
log(exit ramp AADT)	−0.597	0.604	0.983	0.550
log(crossroad AADT)	0.250	0.621	0.167	1.284
Area type				
Rural	Reference			
Urban	1.023	0.499	4.185	2.782

The results also indicate that the exit ramp AADT reduces the chance of WW entry (OR = 0.737). This eventually means that an increase in exit ramp AADT is associated with a decrease in the chance of WW entry. This is probably because the presence of traffic on exit ramps reduces driver confusion and helps the motorists on crossroads to identify exit and entrance ramps accurately.

Finally, the odds of WW entry for the interchanges located in urban areas was found to be 3.9 times compared to rural areas. Therefore, the interchanges in urban areas should be given higher priority to implement WW-related safety countermeasures.

Model 2: For Predicting WW Entry at the Exit Ramp Terminals Connected to Two-Lane Crossroads

This model only considered the exit ramp terminals of diamond interchanges connected to two-lane crossroads. The sample size consisted of 55 exit ramp terminals with history of WW entries and 270 exit ramp terminals with no history of WW entry. The procedure for developing this model is similar to that of Model 1. A Chi-square test between the null and final models indicated that the deviance of the final model is statistically significantly lower ($p < .001$) than the null model. A summary of this model is presented in Table 3. The fitted logistic regression equation is

$$\begin{aligned}\mathbf{logit}(p) = \mathbf{ln}\left(\frac{p}{1-p}\right) = & -2.364 - 0.264(X_{\mathrm{IA1}}) - 1.145(X_{\mathrm{IA2}}) + 2.181(X_{\mathrm{CR}}) + \\ & 1.153(X_{\mathrm{WWSD1}}) + 1.460(X_{\mathrm{WWSD2}}) + 1.285(X_{\mathrm{WWSD3}}) + 1.374(X_{\mathrm{WWSD4}}) - 0.251\left(X_{\mathrm{Signal}}\right) - \\ & 0.597\left(X_{\mathrm{log(exit\ ramp\ AADT)}}\right) + 0.250\left(X_{\mathrm{log(crossroad\ AADT)}}\right) + 1.023(X_{\mathrm{AT}})\end{aligned} \quad (4)$$

Model 3: For Predicting WW Entry at the Exit Ramp Terminals Connected to Multilane Crossroads

The sample size for predicting WW entry at exit ramp terminals connected to multilane crossroads consisted of 73 exit ramp terminals with history and 158 exit ramp terminals with no history of WW entries. The procedure for developing this model is identical to that of Model 1. A Chi-square test between the null and final models indicated that the deviance of the final model is statistically significantly lower ($p < .001$) than the null model. A summary of this model is presented in Table 4. The fitted logistic regression equation, derived from this model, is

Table 4. Summary of Firth's Logistic Regression Model for Exit Ramp Terminals Connected to Multilane Crossroads (Model 3)

Variables	Parameter estimates	Estimated standard error	Chi-square statistics	OR
Intercept		−6.626	1.381	13.144
Intersection angle				
Right	Reference			
Acute	0.864	0.320	0.659	2.373
Obtuse	−0.036	0.268	5.823	0.965
Median on crossroad				
Non-traversable	Reference			
Traversable	1.292	0.304	0.597	3.640
Is corner radius tangent to crossroad edge?				
No	Reference			
Yes	1.162	0.303	20.373	3.196
Distance of WW sign from crossroad				
200 ft. and less	Reference			
201–300 ft.	1.429	0.307	18.600	4.175
301–400 ft.	1.734	0.388	17.592	5.663
401–500 ft.	1.270	0.394	13.344	3.561
More than 500 ft.	1.427	0.403	15.325	4.166
Signalized/unsignalized?				
Unsignalized	Reference			
Signalized	−1.450	0.372	0.687	0.235
log(exit ramp AADT)	−0.560	0.400	1.066	0.571
log(crossroad AADT)	1.365	0.435	3.883	3.916
Area type				
Rural	Reference			
Urban	1.899	0.357	15.022	6.679

$$\begin{aligned}\mathbf{logit}(p) = \mathbf{ln}\left(\frac{p}{1-p}\right) = & -6.626 + 0.864(X_{IA1}) - 0.036(X_{IA2}) + 1.292(X_{MC}) + 1.162(X_{CR}) + \\ & 1.429(X_{WWSD1}) + 1.734(X_{WWSD2}) + 1.270(X_{WWSD3}) + 1.427(X_{WWSD4}) - 1.450(X_{Signal}) - \\ & 0.560(X_{log(exit\ ramp\ AADT)}) + 1.365(X_{log(crossroa\ AADT)}) + 1.899(X_{AT})\end{aligned} \tag{5}$$

Comparison between Model 2 and Model 3

The results of Model 2 and Model 3 can be interpreted in a similar way to those of Model 1. However, it was worth comparing the results of Model 2 and Model 3 to investigate if a particular geometric feature or TCD is more important for the exit ramp terminal connected to a two-lane crossroad than when they meet a multilane crossroad. A comparison between Model 2 and Model 3 reveals that the acute angle intersection dramatically increases the risk of WWD when connected to multilane crossroad compared with two-lane crossroad. The crossroad median was not found to be a significant predictor for two-lane crossroad, although the traversable median increased the odds of WW entry by 3.64 times at exit ramps connected to multilane crossroads. The non-angular connection increased the odds of WWD by 8.855 times for two-lane compared with 3.196 times for multilane crossroads. The distance of WW sign from crossroad had similar effects on the probability of WWD at an exit ramp connected to two-lane or multilane crossroads. The signalized intersection of multilane crossroad and exit ramp is more effective in reducing the chance of WWD (OR = 0.235). Although exit ramp AADT had similar effects for two-lane and multilane crossroad, the crossroad AADT was found to be associated with higher chance of WWD for multilane crossroad (OR = 3.916). Finally, the multilane crossroad and exit ramp intersection in urban areas had higher odds of WWD (OR = 6.679) compared with two-lane crossroads. These results are likely to help transportation agencies to identify which parameters should be given more importance when they apply countermeasures at the exit ramp terminals connected to two-lane or multilane crossroad.

Conclusions and Recommendations

Although full diamond interchanges are less susceptible to WWD compared with partial cloverleaf or trumpet interchanges, the initial entry point of a large portion of WWD crashes can be attributed to the exit ramp terminals of diamond interchanges because they are most widely used nationwide. However, they often draw comparatively less attention when it comes to WWD at certain interchange types. To fill this gap, this study developed mathematical models to predict the risk of WW entry at exit ramp terminals of full diamond interchanges based on geometric design features, usage of TCDs, AADT on exit ramp and connecting crossroad, and area type. Transportation authorities are often challenged by the decision-making process for allocating the limited funding to the best possible use in improving the transportation infrastructure within their jurisdiction. The models developed in this study can be useful tools to help them assess the risk of WW entry on widely used full diamond interchanges, identify and prioritize the best locations for applying countermeasures, and eventually reduce WWD crashes in their jurisdiction.

In this study, some specific characteristics have been identified that contribute to the probability of WW entry at the exit ramp terminals of full diamond interchanges. The following recommendations are drawn based on these results:

1. Although a right angle is recommended by the *AASHTO Green Book* for connecting exit ramps to crossroads, the results show that an obtuse-angle connection was found to have the least possibility of WW entry, as it makes the WW right-turning maneuver difficult. Therefore, the connection between crossroads and exit ramps of full diamond interchanges are recommended to be an obtuse angle for reducing WWD. Future research can focus on determining the type of angle which results in fewest overall crashes.
2. The *AASHTO Green Book* recommends using a non-traversable median on the crossroad and angular connection between the left edge of exit ramp and right edge of crossroad to deter WW entry. The results of this study also show that these geometric features are effective in reducing the chance of WW entry and therefore support the recommendation made by the *AASHTO Green Book*.
3. An obtuse-angle intersection and an angular connection between the left edge of exit ramp and right edge of crossroad makes the right-turning WW maneuver difficult, while a non-traversable crossroad median makes the left-turning WW maneuver less likely. Therefore, a combination of these geometric features is likely to ensure the least possibility of WW entry.
4. Although MUTCD recommends using at least one WW sign on exit ramps, the placement of this sign along the exit ramp is not specified. Based on the results, we recommend that the first WW sign should be used within 200 feet from the crossroads, so that these signs are clearly visible to the motorists on the crossroad.
5. Signalized exit ramp terminals have lower chance of WW entry, as they provide more regulated traffic flow. Therefore, the probability of WW entry can be considered as a supplement to the MUTCD traffic signal warrants to justify the application of signals at the exit ramp terminals. However, further research is necessary to establish proper guidance for incorporating the probability of WW entry as a supplement to the MUTCD traffic signal warrants.
6. The results showed that the locations with low exit ramp AADT and high crossroad AADT are more prone to WW entries. Therefore, such locations should be given higher priority for implementing safety countermeasures.
7. The interchanges in urban areas should be given higher priority for implementing safety countermeasures.

A relatively small sample size from only two states has been used in this study. Small sample size has been a common problem for many WWD studies due to the rare nature of these crashes and limitations in data collection practices. Nonetheless, the future studies can explore the proposed method for a larger sample size collected from a wide range of transportation agencies to make the proposed models more robust and widely applicable.

Acknowledgments

This paper is based on two previous research projects sponsored by Alabama Department of Transportation (ALDOT) and Illinois Center for Transportation (ICT).

Author Contributions

The authors confirm contribution to the paper as follows: study conception and design: Huaguo Zhou, Md Atiquzzaman; data collection: Md Atiquzzaman; analysis and interpretation of results: Md Atiquzzaman; draft manuscripts preparation: Md Atiquzzaman, Huaguo Zhou. All Authors reviewed the results and approved the final version of the manuscript.

References

1. National Transportation Safety Board (NTSB). *Highway Special Investigation Report: Wrong-Way Driving*. Publication NTSB/SIR-12/01, 2012.
2. Baratian-Ghorghi, F., H. Zhou, and J. Shaw. Overview of Wrong-Way Driving Fatal Crashes in the United States. *ITE Journal*, Vol. 84, No. 8, 2014, pp. 41–47.
3. Zhou, H., J. Zhao, R. Fries, M. R. Gahrooei, L. Wang, B. Vaughn, K. Bahaaldin, and B. Ayyalasomayajula.

Investigation of Contributing Factors Regarding Wrong-Way Driving on Freeways. Publication FHWA-ICT-12-010. Illinois Center for Transportation, Illinois Department of Transportation (IDOT), 2012.
4. Missouri Department of Transportation (MoDOT). *Interchanges and Intersections.* http://www.modot.org/InterchangesandIntersections.htm. Accessed June 30, 2017.
5. Moler, S. Stop. You Are Going the Wrong Way! *Public Roads*, Vol. 66, No. 2, 2002, pp. 24–29.
6. Braam, A. C. *Wrong-Way Crashes: Statewide Study of Wrong-Way Crashes on Freeways in North Carolina.* Traffic Engineering and Safety System Branch, North Carolina Department of Transportation, 2006.
7. Neuman, T. R., J. J. Nitzel, N. Antonucci, S. Nevill, and W. Stein. *NCHRP Report 500: Guidance for Implementation of the AASHTO Strategic Highway Safety Plan.* Transportation Research Board, Washington, D.C., 2008.
8. Morena, D. A., and T. J. Leix. Where These Drivers Went Wrong. *Public Roads*, Vol. 75, No. 6, 2012.
9. Copelan, J. E. *Prevention of Wrong Way Accidents on Freeways.* Publication FHWA/CA-TE-89-2. California Department of Transportation, California, 1989.
10. Cooner, S. A., A. S. Cothron, and S. E. Ranft. *Countermeasures for Wrong-Way Movement on Freeway: Overview of Project Activities and Findings.* Publication FHWA/TX-04/4128-1. Texas Transportation Institute, Texas, 2004.
11. Lathrop, S. L., T. B. Dick, and K. B. Nolte. Fatal Wrong-Way Collisions on New Mexico's Interstate Highways, 1990–2004. *Journal of Forensic Science*, Vol. 55, No. 2, 2010, pp. 432–437.
12. Pour-Rouholamin, M., H. Zhou, B. Zhang, and R. E. Turochy. Comprehensive Analysis of Wrong-Way Driving Crashes on Alabama Interstates. *Transportation Research Record: Journal of the Transportation Research Board*, 2016. 2601: 50–58.
13. Ponnaluri, R. V. The Odds of Wrong-Way Crashes and Resulting Fatalities: A Comprehensive Analysis. *Accident Analysis and Prevention*, Vol. 88, 2016, pp. 105–116. http://dx.doi.org/10.1016/j.aap.2015.12.012
14. Baratian-Ghorghi, F., H. Zhou, M. Jalayer, and M. Pour-Rouholamin. Prediction of Potential Wrong-Way Entries at Exit Ramp Terminals of Signalized Partial Cloverleaf Interchanges. *Traffic Injury Prevention*, Vol. 16, No. 6, 2014, pp. 599–604. http://dx.doi.org/10.1080/15389588.2014.981651
15. Sandt, A., H. Al-Deek, and J. H. Rogers, Jr. Identifying Wrong-Way Driving Hotspots by Modeling Crash Risk and Analyzing Traffic Management Center Response Times. Presented at 96th Annual Meeting of the Transportation Research Board, Washington, D.C., 2017.
16. Rogers Jr, J. H., H. Al-Deek, A. H. Alomari, E. Gordin, and G. Carrick. Modeling the Risk of Wrong-Way Driving on Freeways and Toll Roads. *Transportation Research Record: Journal of the Transportation Research Board*, 2016. 2554: 166–176.
17. Rogers Jr., J. H., A. Sandt, H. Al-Deek, A. H. Alomari, N. Uddin, E. Gordin, C. D. Santos, J. Renfrow, and G. Carrick. Wrong-Way Driving Multifactor Risk-Based Model for Florida Interstates and Toll Facilities. *Transportation Research Record: Journal of the Transportation Research Board*, 2015. 2484: 119–128.
18. Pour-Rouholamin, M., and H. Zhou. Logistic Model to Predict the Effect of Various Geometric Design Elements on the Probability of Wrong-Way Entries at Partial Cloverleaf Interchanges. Presented at 95th Annual Meeting of the Transportation Research Board, Washington, D.C., 2016.
19. Zhou, H., and M. Pour-Rouholamin. *Guidelines for Reducing Wrong-Way Crashes on Freeways.* Publication FHWA-ICT-14-010. Illinois Center for Transportation, Illinois Department of Transportation, 2014.
20. American Association of State Highway and Transportation Officials (AASHTO). *A Policy on Geometric Design of Highways and Streets (Green Book).* AASHTO, Washington, D.C., 2011.
21. Eyler, D. R. Arterial Interchange. *Transportation Research Record: Journal of the Transportation Research Board*, 2005. 1912: 65–71.
22. California Department of Transportation. *Highway Design Manual.* California Department of Transportation, Sacramento, CA, 2014.
23. Washington State Department of Transportation. *Design Manual.* Washington State Department of Transportation, Olympia, WA, 2013.
24. Al-Ghamdi, A. S. Using Logistic Regression to Estimate the Influence of Accident Factors on Accident Severity. *Accident Analysis and Prevention*, Vol. 34, No. 6, 2002, pp. 729–741. http://dx.doi.org/10.1016/S0001-4575(01)00073-2
25. Yan, X., E. Radwan, and M. Abdel-Aty. Characteristics of Rear-End Accidents at Signalized Intersections Using Multiple Logistic Regression Model. *Accident Analysis and Prevention*, Vol. 37, No. 6, 2005, pp. 983–995. http://dx.doi.org/10.1016/j.aap.2005.05.001
26. Sarkar, S., R. Tay, and J. Hunt. Logistic Regression Model of Risk of Fatality in Vehicle-Pedestrian Crashes on National Highways in Bangladesh. *Transportation Research Record: Journal of the Transportation Research Board*, 2011. 2264: 128–137.
27. Qin, X., K. Wang, and C. Cutler. Logistic Regression Models of the Safety of Large Trucks. *Transportation Research Record: Journal of the Transportation Research Board*, 2013. 2392: 1–10.
28. Torrão, G., M. Coelho, and N. Rouphail. Modeling the Impact of Subject and Opponent Vehicles on Crash Severity in Two-Vehicle Collisions. *Transportation Research Record: Journal of the Transportation Research Board*, 2014. 2432: 53–64.
29. Yu, C. How Differences in Roadways Affect School Travel Safety. *Journal of the American Planning Association*, Vol. 81, No. 3, 2015, pp. 203–220. http://dx.doi.org/10.1080/01944363.2015.1080599
30. Firth, D. Bias Reduction of Maximum Likelihood Estimates. *Biometrika*, Vol. 80, No. 1, 1993, pp. 27–38. http://dx.doi.org/10.1093/biomet/80.1.27
31. King, G., and L. Zeng. Logistic Regression in Rare Events Data. *Political Analysis*, Vol. 9, No. 2, 2001, pp. 137–163. http://dx.doi.org/10.1093/oxfordjournals.pan.a004868

32. Heinze, G., and M. Schemper. A Solution to the Problem of Separation in Logistic Regression. *Statistics in Medicine*, Vol. 21, No. 16, 2002, pp. 2409–2419. http://dx.doi.org/10.1002/sim.1047
33. Allison, P. D. Convergence Failures in Logistic Regression. *Statistics and Data Analysis, SAS Global Forum*, Vol. 360, 2008, pp. 1–11.
34. Federal Highway Administration (FHWA). *Manual on Uniform Traffic Control Devices for Streets and Highways.* FHWA, Washington, D.C., 2009.

The Standing Committee on Operational Effects of Geometrics (AHB65) peer-reviewed this paper (18-05085).

Article

TRR
JOURNAL OF THE TRANSPORTATION RESEARCH BOARD

Transportation Research Record
2018, Vol. 2672(17) 48–60

DOI: 10.1177/0361198118778931
journals.sagepub.com/home/trr
SAGE

Drivers' Speeding Behavior on Expressway Curves: Exploring the Effect of Curve Radius and Desired Speed

Charitha Dias[1], Takashi Oguchi[1], and Kasun Wimalasena[2]

Abstract

Speed profiles can be considered as a key input for assessing safety, comfort and efficiency of highway or expressway segments. Therefore, understanding drivers' speeding behavior, particularly on expressway curve sections, is important. Most previous studies have modeled the speed on highway curve sections mainly as constant or a piecewise linear profile. Such approaches may not realistically represent the properties of speed and acceleration behavior. Furthermore, mechanisms underlying the speeding behavior through curve sections have not been comprehensively studied. In this study, the minimum-jerk concept, which was originally applied in neuroscience and robotics domains, is utilized to explore drivers' speeding behavior on expressway curve sections. GPS-based naturalistic driving data of vehicles travelling on the Tomei expressway in Japan under free-flow conditions were used to explore the applicability and validity of the proposed approach. How the proposed approach can be used to evaluate the effect of horizontal geometry and desired driving speed on drivers' speeding and acceleration behavior on expressway curve sections is also discussed. The findings of this study could be useful in modeling speed and acceleration choice behaviors on highway curve sections which could potentially be applied in highway design consistency evaluations. Furthermore, the outputs of this study may be useful in other advanced applications, such as modeling and visualizing realistic vehicle movements in driving simulators and virtual reality applications and trajectory planning of autonomous vehicles.

The main objective of highway geometric elements (i.e., horizontal and vertical alignments) is to facilitate safe, efficient and comfortable traffic operations (*1*, *2*). To assess such properties (or design consistency in a broad sense) knowledge of speed variation or speed profiles along highway or expressway segments is important. Estimating the speed difference between tangent and horizontal curve segments along a roadway is considered to be a popular criteria for evaluating design consistency (*3*, *4*). Thus, accuracy and reliability of such estimates largely depends on the accuracy of the predicted speed profiles. Nevertheless, most previous studies have simplified speed profiles on curve sections by assuming constant or piecewise linear patterns.

Early studies, which focused on estimating speeds on tangent and curve sections, used simplified methods such as stop watches (*5*, *6*) and traffic counters or classifiers (*7*). Average speed of a sample of vehicles passing a road section of a certain length can be estimated with such data. Radar or Lidar guns have also been utilized in previous studies to measure spot speeds of a sample of vehicles passing given points, for example, particular points on the tangent and the circular curve (*7*, *8*). A detailed review that discusses operating speed models based on such data sources was previously published by Misaghi and Hassan (*9*). These models assume a constant speed along the entire horizontal curve or a portion of it, which is a major limitation. Therefore, drivers' speed changes or acceleration-deceleration behaviors inside curve sections are not accurately represented. To overcome the limitations of such models based on spot speed measurements, new data collection methods, which output continuous speed profiles, became popular in later years. Data collected through vehicles equipped with GPS devices and driving simulators are famous among such approaches. For example, Zuriaga et al., Cafiso and Cerni and Montella et al. modeled operating speed profiles on rural roads utilizing data collected through instrumented vehicles with GPS devices (*10–12*). Other than operating speeds, acceleration/deceleration behaviors have also been studied with such data by previous authors (*13*, *14*). Driving simulators have been widely used in recent years to collect data for modeling continuous speed profiles (*15*, *16*). Other studies, which modeled speeds on curves and tangents using data collected

[1]Institute of Industrial Science, The University of Tokyo, Tokyo, Japan
[2]Department of Civil and Environmental Engineering, University of Ruhuna, Galle, Sri Lanka

Corresponding Author:
Address correspondence to Charitha Dias: cdias@iis.u-tokyo.ac.jp

through instrumented vehicles and driving simulators, are listed in Montella et al. (*12*). Although these data sources provide continuous speed profiles for understanding speed behavior on highway curve sections there still exists critical shortcomings. Volunteers or subject drivers, who are participating in instrumented vehicle and driving simulator experiments, know that they are being monitored. Furthermore, there are realism and validity issues in driving simulators. Thus, realistic driving behaviors may not be captured and that results in bias in collected data. Therefore, naturalistic driving data, which provide bias-free continuous speed data, may be more suitable for understanding drivers' speeding behavior on highway curve sections (*17*). Almost all of these previous studies, which were based on continuous speed data, modeled the speed and acceleration profiles on curve sections as piecewise linear functions using regression based methods. Thus, realistic properties of speed and acceleration profiles have not been represented. Furthermore, mechanisms underlying drivers' speeding or accelerating/decelerating behaviors (e.g., on curve sections) have not been comprehensively understood. Realistic representation of speed and acceleration behaviors in explanatory models could be important not only in geometric design related applications but in other advanced applications such as 3-D visualization of vehicle movements in driving simulators, virtual reality applications and trajectory planning for autonomous vehicles as well.

Bosetti et al. (*18*) modeled curve negotiation behavior using the two-thirds law, a principle which was originally used to describe characteristics of human locomotion, and motor primitives (*19*). Dias et al. (*20*) used the minimum-jerk principle (*21*) to model speed and acceleration profiles of turning vehicles at signalized intersections. These approaches have been validated with empirical trajectories. Gleaning insights from such studies, the objective of this this paper is to explore the applicability of the minimum jerk principle to model drivers' speeding behavior on expressway curve sections. GPS-based naturalistic driving data for vehicles travelling on an expressway in Japan are used to test the proposed modeling approach.

This paper is structured as follows: the next section will discuss the details of the data sources and modeling methods. Geometric characteristics of the curve sections utilized in this study are also described. This is followed by a discussion on results. Finally, the conclusions are presented.

Data and Methods

Data Sources

This study is based on free-flow vehicle trajectory data collected on the Tomei Expressway which is a toll road operated by the Central Nippon Expressway Company of Japan. This expressway links Tokyo and Nagoya cities and a map is shown in Figure 1.

Trajectory data of vehicles recorded from January 2013 to September 2014 were extracted from a database (of big data) for 12 horizontal curve sections. The geometric characteristics of the curve sections considered in this paper are summarized in Table 1 and approximate locations of the considered curve sections are shown in Figure 1. The Tomei expressway, of which the length is approximately 347 km, consists of 2- and 3-lane sections. Furthermore, the speed limit of some sections is 80 km/h whereas that of other sections is 100 km/h. In this study, only the data for the sections of which the speed limit is 100 km/h were used.

The database, from which the trajectories were extracted for this study, was obtained from the ETC 2.0 system which is an on-board navigation system that mainly provides area-wide traffic congestion information (*22*). This system records the GPS position, GPS time and the speed of vehicles and stores it temporarily until the vehicle encounters a roadside ITS infrastructure (ITS spot) to upload those records and to receive updated service information via bi-directional communication. Other than the data extracted from the ETC 2.0 database, loop detector data, weather data and incident data were also utilized to estimate the free-flow state and to extract free-flow vehicle trajectories. A detailed description of these data sources and steps followed to filter the data to extract free-flow vehicle trajectories can be found in Wimalasena et al. (*23*).

The extracted data, which were used in this study, consists of the GPS time, GPS location and speed recorded for each vehicle (with encoded vehicle identification) at 100–200 m intervals. A path reconstructed with a sample of 70 free flow trajectories and corresponding speed profiles for the same trajectory sample for one curve segment is shown in Figure 2. As can be understood from this figure, the free-flow speeds largely vary from approximately 70 km/h to 140 km/h. The average (±standard deviation) speed for the particular road section shown in Figure 2 was estimated as 104 (±14.4) km/h, which is very close to the speed limit 100 km/h.

It is hard to obtain a generic speed or acceleration profile, which describes drivers' speeding behaviors, from these empirical data. Thus, these data were utilized to obtain trajectories (theoretically) on curve segments based on the minimum jerk concept. Details of the theoretical background of the modeling principle is described in the next section.

Minimum-Jerk Concept

The minimum-jerk concept was initially utilized in the neuroscience domain to study optimality characteristics of skilled human arm movements. Flash and Hogan verified that smoothness of a skilled human planar arm movements (e.g., reaching, writing and drawing tasks) can be evaluated as a function of jerk (jerk is defined as the time derivative of acceleration) (*21*). They verified the objective function that should be minimized to obtain the smoothest trajectory for a moving hand from an initial position to a final position in a

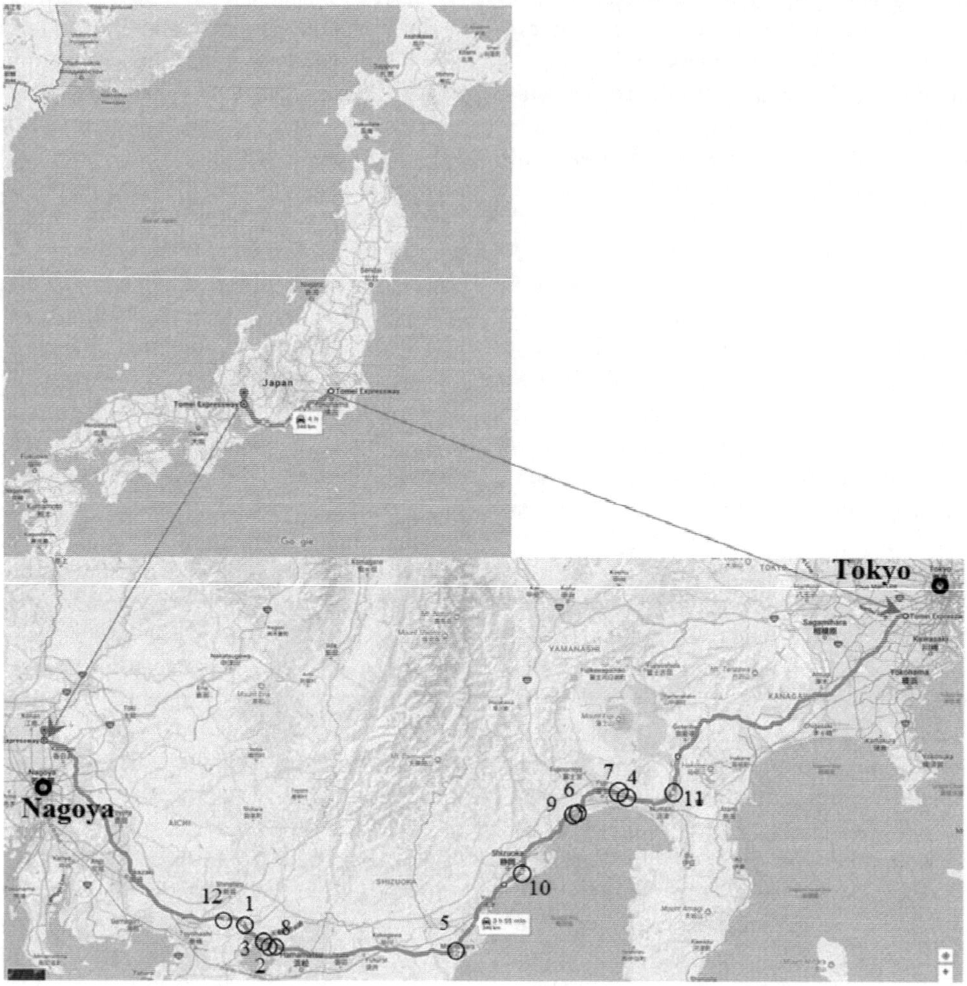

Figure 1. Map of Tomei Expressway and approximate locations of considered curve sections (source: Google Maps).

given time interval T ($T = t_f - t_i$) is the time integration of the square of jerk that was expressed as:

$$J = \frac{1}{2}\int_{t_i}^{t_f}\left(\left(\frac{d^3x}{dt^3}\right)^2 + \left(\frac{d^3y}{dt^3}\right)^2\right)dt \quad (1)$$

The solution for this minimization problem can be obtained as proved in Flash and Hogan (*21*) as a set of fifth order polynomials of time, with a_j and b_j ($j = \{0, ..., 5\}$) constants, as follows:

$$x(t) = a_0 + a_1t + a_2t^2 + a_3t^3 + a_4t^4 + a_5t^5 \quad (2)$$

$$y(t) = b_0 + b_1t + b_2t^2 + b_3t^3 + b_4t^4 + b_5t^5 \quad (3)$$

The minimum-jerk concept has been applied to understand characteristics of human goal-oriented and turning maneuvers (*24*, *25*). Furthermore, this concept has been applied in movement planning problems in robot limbs (*26*, *27*) and motion planning and control problems in autonomous vehicles (*28*, *29*). Car following behavior was also described with the minimum jerk concept assuming that the approaching behavior of the following car (toward the vehicle in front) matches with skilled reaching movements of the human arm (*30*). Dias et al. explained that the trajectories of turning vehicles under free-flow conditions at intersections can also be described with the minimum jerk principle (*20*). These studies, for example mainly Dias et al., Bianco et al., Shamir and Hiraoka et al. confirm the applicability of the minimum jerk principle to understand and model drivers' speed and acceleration behaviors under free-flow conditions (*20*, *28*–*30*).

Boundary Conditions

The main goal of this study it to obtain speed and acceleration profiles for each considered curve segments utilizing the minimum-jerk principle. It should be noted that the first and

Table 1. Geometric Characteristics of the Considered Curve Sections

Curve ID	KP start (km)	KP end (km)	Length (m)	Type	Radius (m)	Gradient (%)
1	256.325	256.488		Tangent	300	1.6
	256.488	256.823	335	Circular curve	550	1.6
	256.823	256.937		Tangent	250	1.6
2	245.748	245.909		Circular curve	400	0.5
	245.909	246.524	615	Circular curve	650	0.5
	246.524	246.909		Tangent	500	0.5
3	248.067	248.196		Tangent	300	−1.1
	248.196	248.475	279	Circular curve	700	−1.1
	248.475	248.623		Tangent	500	−1.1
4	114.689	115.002		Tangent	500	1.6
	115.002	115.041		Circular curve	800	1.6
	115.041	115.329	327	Circular curve	800	−0.3
	115.329	115.582		Tangent	450	−0.3
5	188.724	188.978		Tangent	450	0.9
	188.978	189.214	236	Circular curve	800	0.9
	189.214	189.467		Tangent	450	0.9
6	132.414	132.592		Tangent	400	−0.4
	132.592	132.773	181	Circular curve	900	−0.4
	132.773	132.998		Tangent	450	−0.4
7	115.582	115.785		Tangent	450	−0.3
	115.785	116.043	258	Circular curve	1,000	−0.3
	116.043	116.293		Tangent	500	−0.3
8	244.504	244.706		Tangent	450	−0.5
	244.706	244.748		Circular curve	1,000	−0.5
	244.748	244.955	249	Circular curve	1,000	−1.8
	244.955	245.077		Tangent	350	−1.8
9	132.998	133.167		Tangent	450	−0.4
	133.167	133.581	414	Circular curve	1,200	−0.4
	133.581	133.750		Tangent	450	−0.4
10	157.598	157.600		Tangent	800	−1.4
	157.600	158.139	539	Circular curve	1,300	−1.4
	158.139	158.631		Tangent	800	−1.4
11	98.438	98.839		Tangent	750	0.8
	98.839	99.101		Circular curve	1,400	0.8
	99.101	99.402	563	Circular curve	1,400	−1.3
	99.402	99.752		Tangent	700	−1.3
12	261.928	262.099		Tangent	600	−0.3
	262.099	262.455	356	Circular curve	1,500	−0.3
	262.455	262.641		Tangent	500	−0.3

Note: KP = kilo-post, which indicates the distance measured from the start of the expressway; ID = identity.

second time derivatives of Equations 2 and 3 provides the velocity and acceleration vectors respectively. To solve these equations, 12 boundary conditions, along with the movement time within those boundaries $T\ (= t_f - t_i)$ are required. The x- and y-components of location, velocity and acceleration vectors (in 2-D space) at the initial and final points provide the 12 boundary conditions. For the curve segments considered in this study acceleration data are not available at the point of curvature (PC) and the point of tangent (PT). Furthermore, as the data have been recorded in 100–200 m intervals, it is not practical to estimate accelerations from those data. Thus, the initial and final position of the modeled trajectory were chosen as 100 m before the PC and 100 m after the PT, respectively. Meaning that we consider that drivers start to decelerate at 100 m before the PC and regain their speed at 100 m after the PT.

Altamira et al. demonstrated that vehicles approaching at curves, of which the radii are between 500 m and 1000 m and the gradients are less than 2%, start to decelerate approximately 50–230 m before the PC (*31*). Furthermore, for the vehicles leaving the curve, the point where the acceleration ends was estimated approximately as 150–270 m after the PT.

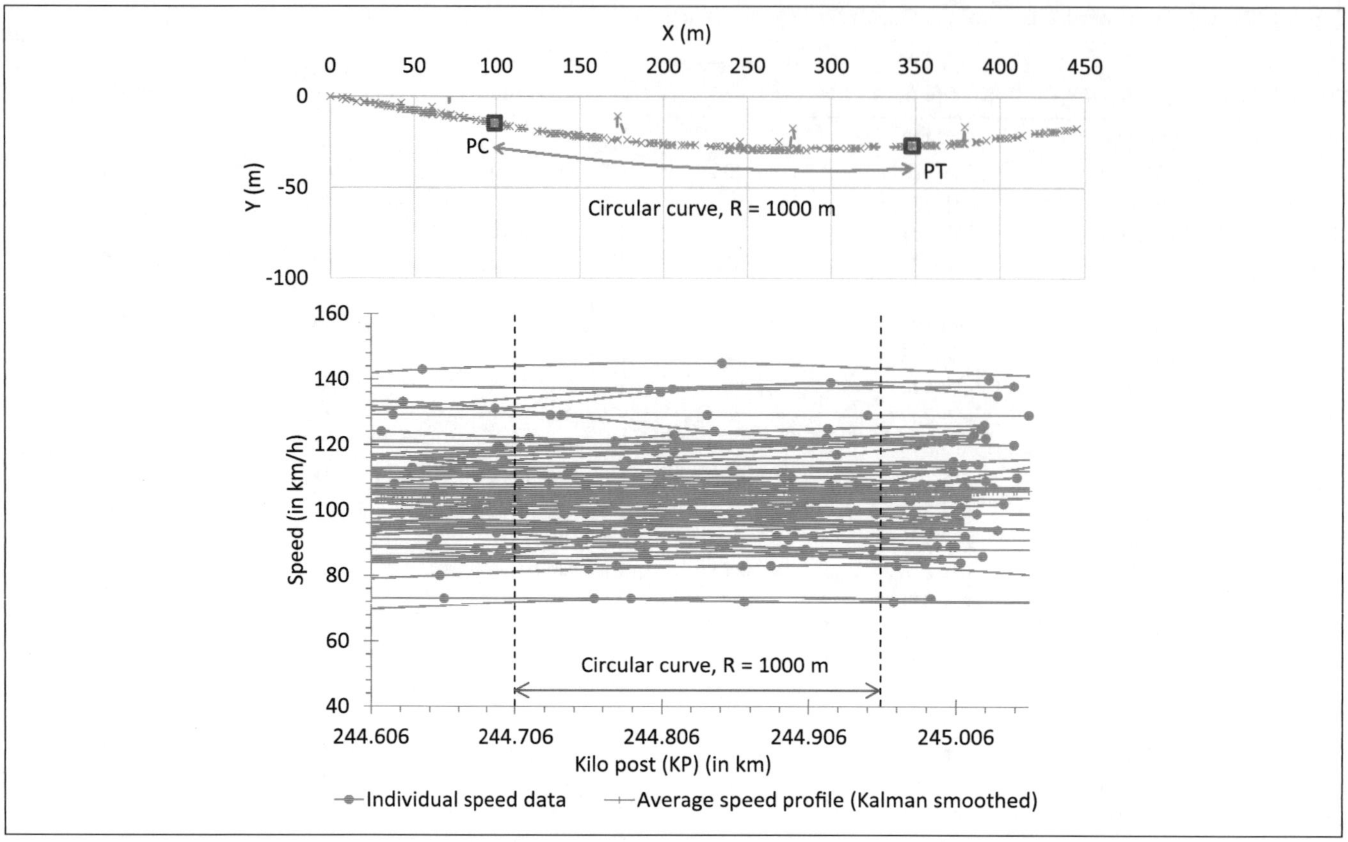

Figure 2. Curve segment reconstructed with GPS points and a sample of free-flow speed profiles for curve ID 8.

However, as indicated by the study of Montella et al., this distance can vary approximately between 50 m and 200 m for curves of which the radii are within 600 m and 800 m (*16*). The study by Pérez Zuriaga et al. described that the deceleration starts approximately 70 m before the beginning of the transition curve for curves of which the radii are less than 519 m (*32*). From these studies, it is clear that the locations of these points could largely be varied by numerous factors, including curve radius and desired speed, and that further studies are needed to tackle such issues. In this study it is assumed that drivers start to decelerate approximately 100 m before the PC and stop accelerating and regain their speed at approximately 100 m after the PT for all the considered 12 curves. Later in the results section the sensitivity of such assumptions is tested. Coordinates of those locations were obtained from the path reconstructed from the GPS location data (see Figure 2) reported on each section. Speeds at those entry and exit points were interpolated for each trajectory using the nearest speed data recorded before and after those points. Movement time between these points ($T = t_f - t_i$) were also estimated for each trajectory using recorded GPS time and location data and speed values before and after the considered starting and ending points (i.e., 100 m before and after the PC and PT). Data from 50–70 randomly chosen free-flow trajectories traversed on the considered sections were used for location, speed and movement time estimations. Boxplots of entry speeds and exit speeds for each of the sections are shown in Figure 3. It can be noted from Figure 3 and Table 2 that the average entry speeds do not vary according to the curve radius. The general expectation is that the vehicles will approach a curve with a larger radius with higher speeds than a curve with a smaller radius. It should be noted that the preceding geometric conditions may have considerable influence on the entry speed (*33*). Nevertheless, it can be noted that the entry and exit speeds for all considered curve sections are not considerably deviated from the speed limit, which is 100 km/h.

Initially, Equations 2 and 3 were solved for the boundary conditions and T for the average trajectory of each curve (i.e., using average entry and exit speeds and average T). The boundary conditions and T used to obtain the average minimum-jerk trajectories are shown in Table 2. Note that the movement times (T) are largely varied mainly because of the differences in curve lengths.

Results

Effect of Curve Radius on Vehicle Trajectories on Curves

The resulting average minimum-jerk paths solved with each set of boundary conditions and T are compared with the corresponding actual paths reconstructed from GPS points of

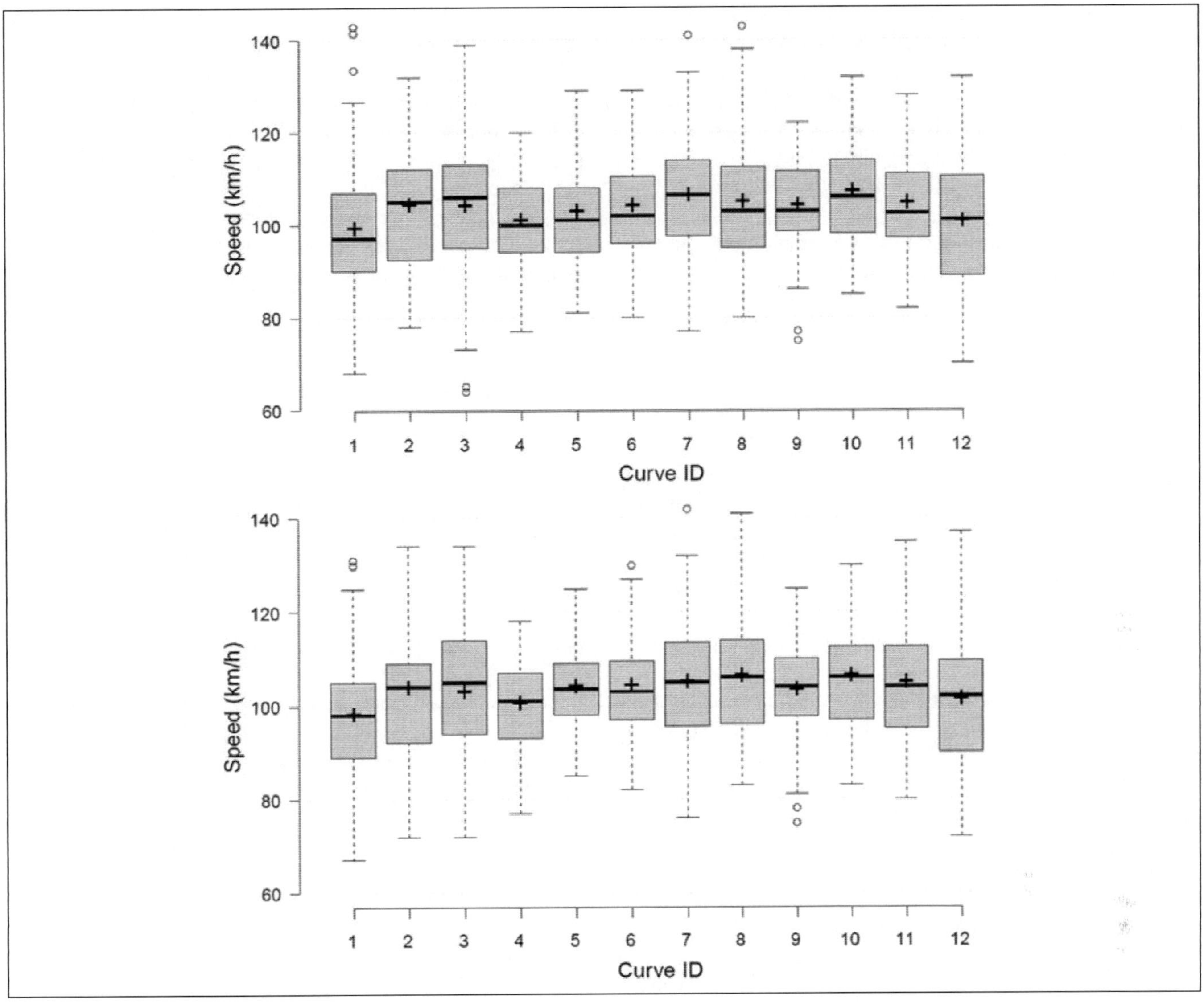

Figure 3. Boxplot of entry (top) and exit (bottom) speeds for each curve section (+ indicates the sample mean, curve IDs are given in Table 1).

Table 2. Boundary Conditions and *T* for Solving Equations 2 and 3 for Average Minimum-Jerk Trajectories

Curve ID	Avg. entry speed ± SD (km/h)	Entry acceleration (m/s^2)	Avg. exit speed ± SD (km/h)	Exit acceleration (m/s^2)	Avg. movement time ± SD (s)
1	98.6 ± 14.8	0	97.6 ± 12.8	0	20.2 ± 2.7
2	103.7 ± 13.0	0	103.2 ± 14.4	0	28.7 ± 3.7
3	103.6 ± 17.5	0	102.5 ± 15.5	0	17.0 ± 3.0
4	100.5 ± 10.2	0	100.0 ± 10.1	0	19.6 ± 2.9
5	102.3 ± 10.9	0	103.5 ± 9.1	0	15.4 ± 1.5
6	103.6 ± 11.3	0	103.7 ± 11.0	0	13.4 ± 1.4
7	105.5 ± 12.3	0	104.4 ± 13.1	0	15.7 ± 2.1
8	104.3 ± 14.2	0	105.8 ± 14.2	0	15.6 ± 2.0
9	103.6 ± 11.0	0	102.8 ± 11.3	0	21.5 ± 2.5
10	106.6 ± 11.3	0	105.8 ± 10.6	0	24.8 ± 2.5
11	104.0 ± 11.1	0	104.2 ± 12.4	0	26.5 ± 3.1
12	100.2 ± 14.7	0	100.7 ± 14.8	0	20.1 ± 3.0

Note: Avg. = average; SD = standard deviation; ID = identity.

traversed vehicles in Figure 4. The maximum errors between the estimated and actual paths (in Figure 4) ranged from 0.92 to 6.59 m. The average of the maximum location errors for all curves were estimated as 2.84 m. Observing Figure 4 and the error statistics it can be understood that the curved path can be estimated with the minimum jerk approach with a good accuracy even though only the geometric information of the two boundaries were used. These paths are the minimum jerk paths, or in other terms "smoothness maximized paths". To maximize the smoothness (by minimizing the jerk) drivers have to choose and adjust speeds and accelerations appropriately along the curves.

Speed, acceleration and jerk profiles for each curve segments were estimated from the first, second and third time derivatives of Equations 2 and 3 respectively. Those profiles are shown in Figure 5. As the entry, exit speeds and curve lengths were different, normalized speed profiles were compared for those curves as depicted in Figure 5*a*. Speeds were normalized with respect to the entry speed at each segment and distance along the curve segment was normalized with respect to the curve length (i.e., length from PC to PT). It can be understood from Figure 5*a* that the speeds were not reduced by more than 5% for all considered curve sections of which the radii are more than 550 m. Furthermore, the location of minimum speed can be either on the first half or the second half of the circular curve. The location of minimum speed could be varied depending on the entry and exit speeds other than the curve radius. It should further be noted that the radii of the curves before and after the considered curve may have considerable impact on the entry and exit speeds and therefore on the speed profiles on the considered curve portion. Observing acceleration profiles in Figure 5*b* and the maximum acceleration/deceleration values in Figure 6*b*, it can be understood that the maximum deceleration and maximum acceleration values generally increase with decreasing curve radius. Furthermore, it is evident that even for the smoothest trajectories the deceleration and acceleration could reach up to 0.2 m/s^2. These values may further be increased for curves with smaller radii. Additionally, it can be noted that for some curves the locations of the maximum deceleration and acceleration lie outside the circular portion. As described in the data and methods section, acceleration values at 100 m before the PC and after the PT points were assumed to be zero. Such assumptions could have an impact on the speed and acceleration profiles. Figure 5*c* compares the jerk profiles for the 12 considered curves and those profiles display similar trends, for example, increase in jerk followed by a decrease, within the circular curve section.

The percentage maximum speed reduction within the considered curve segments are compared in Figure 6*a*. It can be understood that the speed reduction is generally decreased with increasing curve radius. However, some exceptions can also be clearly visible (e.g., for R = 900 m and R = 1300 m curves). Such deviations could be because of the effect of gradient and properties of the tangent sections (tangents were clothoid curves of different dimensions for all considered sections). Furthermore, Wimalasena et al. explained that the effect of gradient is more prominent compared to the radius of the curve (for curve radii greater than 550 m) (*23*).

The jerk cost (Equation 1) within the curve section for the minimum jerk trajectory are compared for the considered 12 sections in Figure 6*c*. It is clear that in general the jerk cost tends to decrease with increasing curve radius. That is, trajectories on curves with larger radii are relatively smoother compared to the trajectories on curves with smaller radii.

Effect of Desired Entry Speed and Entry Acceleration on Speed and Acceleration Behaviors Inside the Curve

Desired driving speed and entry acceleration could largely affect the speed and acceleration profiles of minimum-jerk trajectories. Such effects are discussed in this section. Free flow trajectories for 2 different speed levels, for example, approximately 100 km/h (to represent drivers keeping to the speed limit) and approximately 120 km/h (to represent high speed drivers), were extracted (10 free-flow vehicles for each speed category) for the curve ID 5 with R = 800 m (see Table 1 for other geometric details). The entry and exit speeds for these speed levels were approximately 100 km/h and 123 km/h. Estimated (based on the same method described in the section on the minimum jerk concept) movement times for 100 km/h and 123 km/h speeds were 15.6 s and 12.9 s respectively.

Similar to the first section in the results, keeping the deceleration starting point and acceleration end point as 100 m, initially Equations 2 and 3 were solved for different speeds (100 km/h and 123 km/h). For these calculations the entry and exit accelerations were assumed as 0 m/s^2. The resulting paths, speed, acceleration and jerk profiles are depicted in Figure 7, *a–d*. The speed, acceleration and jerk profiles appear to be symmetrical with respect to the mid of the circular curve as the entry and exit speeds, and accelerations were kept similar. However, this is not always the case, as can be observed in Figure 5 and speed profiles in previous studies (*16*). It is clear from Figure 7*a* that the modeled vehicle paths are not deviated from each other and from the actual path. This observation indicates that the geometrically smooth path (i.e., the combination of the tangent and circular curve segments represented by the GPS points) provides a smooth (jerk-minimized) transition for drivers with different speeds. Speed and acceleration profiles (Figure 7, *b* and *c*) explain that the speed reduction and acceleration/deceleration tend to be larger for drivers with higher desired speeds. That is, drivers with higher desired speeds (or entry speeds) tend to decelerate and accelerate more to negotiate the curve smoothly compared to slower drivers. Jerk cost is also higher and therefore the relative smoothness is lower for higher desired speeds compared to lower desired speeds as can be

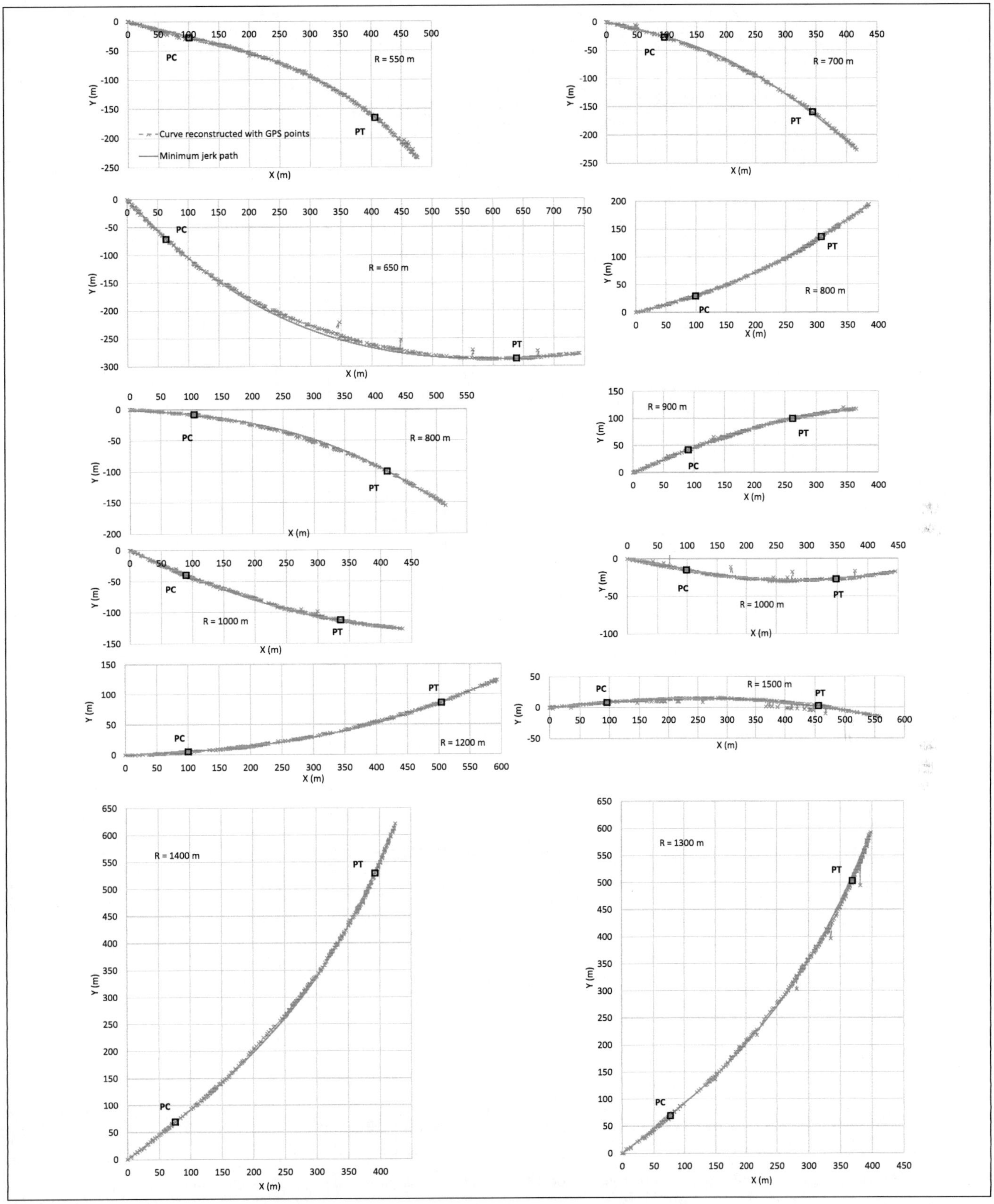

Figure 4. Comparison of minimum-jerk and actual free-flow vehicle paths for 12 considered curves. Blue = path reconstructed from recorded GPS points by vehicles; orange = path obtained with the minimum-jerk principle.

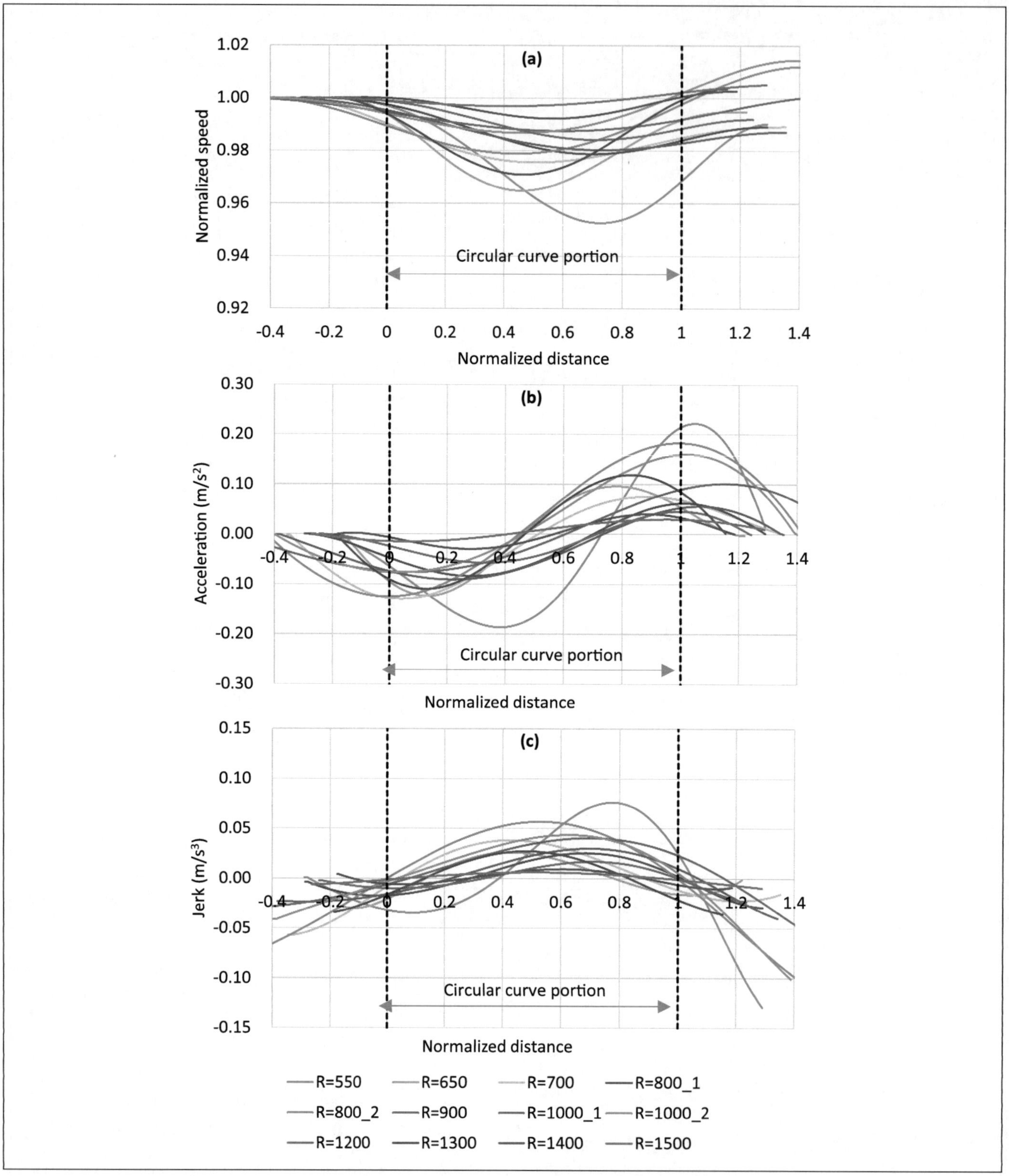

Figure 5. *(a)* Normalized minimum-jerk speed profiles; *(b)* acceleration profiles; *(c)* jerk profiles for different curves.

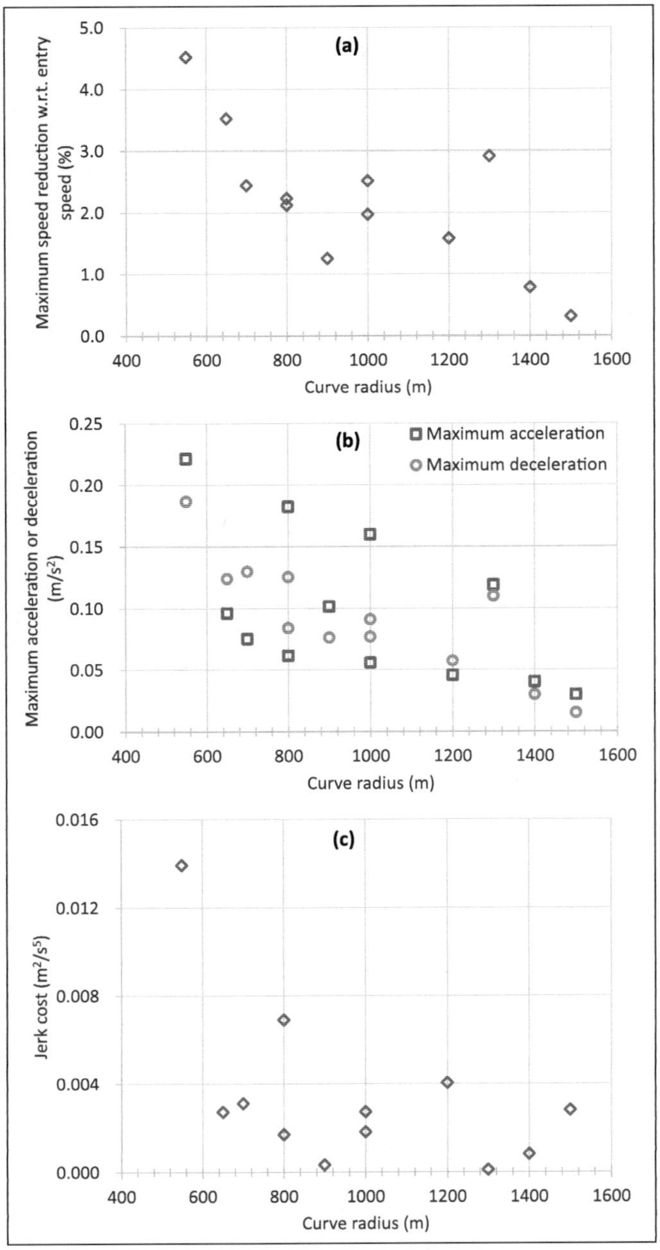

Figure 6. Comparison of (*a*) percentage maximum speed reduction, (*b*) maximum acceleration/deceleration, and (*c*) jerk cost (Equation 1) for different curves.

explained with Figure 7*d*. Such variations in speed and acceleration patterns are adequately captured by the approach tested in this study.

The sensitivity of the trajectories to the entry and exit accelerations was also tested. In these calculations how speed, acceleration and jerk profiles are varied based on the entry deceleration and exit acceleration (i.e., assuming 2 entry deceleration values as -0.1 m/s^2, -0.2 m/s^2 and 2 exit acceleration values as 0.1 m/s^2, 0.2 m/s^2 at 100 m before and after PC and PT) were explored. That is, the deceleration starting point and acceleration end point were assumed to be larger than 100 m. Note that, as indicated in previous studies, these distances could be different for different curve radii and different speed levels (*16*, *31*, *32*), however, no solid evidence was available to choose representative values for the considered speed levels in this study. Keeping such limitations in mind, Equations 2 and 3 were solved for such boundary conditions (for 123 km/h speed level and assuming movement times do not change) and resulting speed, acceleration and jerk profiles were compared as shown in Figure 7, *e–g*. As can be understood from Figure 7e speed profiles do not display a large deviation, however, acceleration (Figure 7*f*) and jerk (Figure 7*g*) profiles show a relatively larger sensitivity. Furthermore, as can be described from Figure 7*g*, the relative smoothness (as a measure of jerk) seems to improve when the entry deceleration is increasing, indicating that drivers, who plan the curve negotiation beforehand, could negotiate the curve 'smoothly'.

Conclusions

Knowledge of drivers' speeding behaviors on highway curve sections is important to evaluate safety, comfort and efficiency of highway and expressway systems. In this study, applicability of the minimum jerk concept to model drivers' speed and acceleration behavior in response to the radius of the curve and desired speed was explored. Qualitative and quantitative comparisons of modeled vehicle paths and corresponding actual paths (reconstructed from GPS positions of traversed vehicles) suggested that the minimum-jerk concept could be applied to reproduce vehicle trajectories on expressway curve sections. Furthermore, the speed, acceleration and jerk profiles were also compared for 12 curve sections. The results suggest that the effect of curve radius on speed behaviors generally increases with decreasing radius. This finding is consistent with results obtained in previous studies. Furthermore, percentage speed reduction, maximum acceleration and maximum deceleration were observed to be increasing with increased desired speed. Such variations can also be captured realistically by the minimum jerk approach explored in this study. Thus, the findings in this study could potentially be utilized in various applications in which information on speed and acceleration behaviors are required.

In this study, the deceleration starting point and acceleration end point were approximated for all curves and for all speed levels. Furthermore, the effect of vertical geometry was ignored. Such approximations may have an effect on results obtained in this study. Thus, validity of such assumptions and sensitivity of those values should be confirmed by future studies using additional data with higher resolution. Additionally, the sensitivity of boundary conditions and movement times should also be tested.

To obtain minimum jerk trajectories, only the geometric information of the boundaries was used. However, geometrical information of the whole curved path is generally known. Such information could also be added as an additional

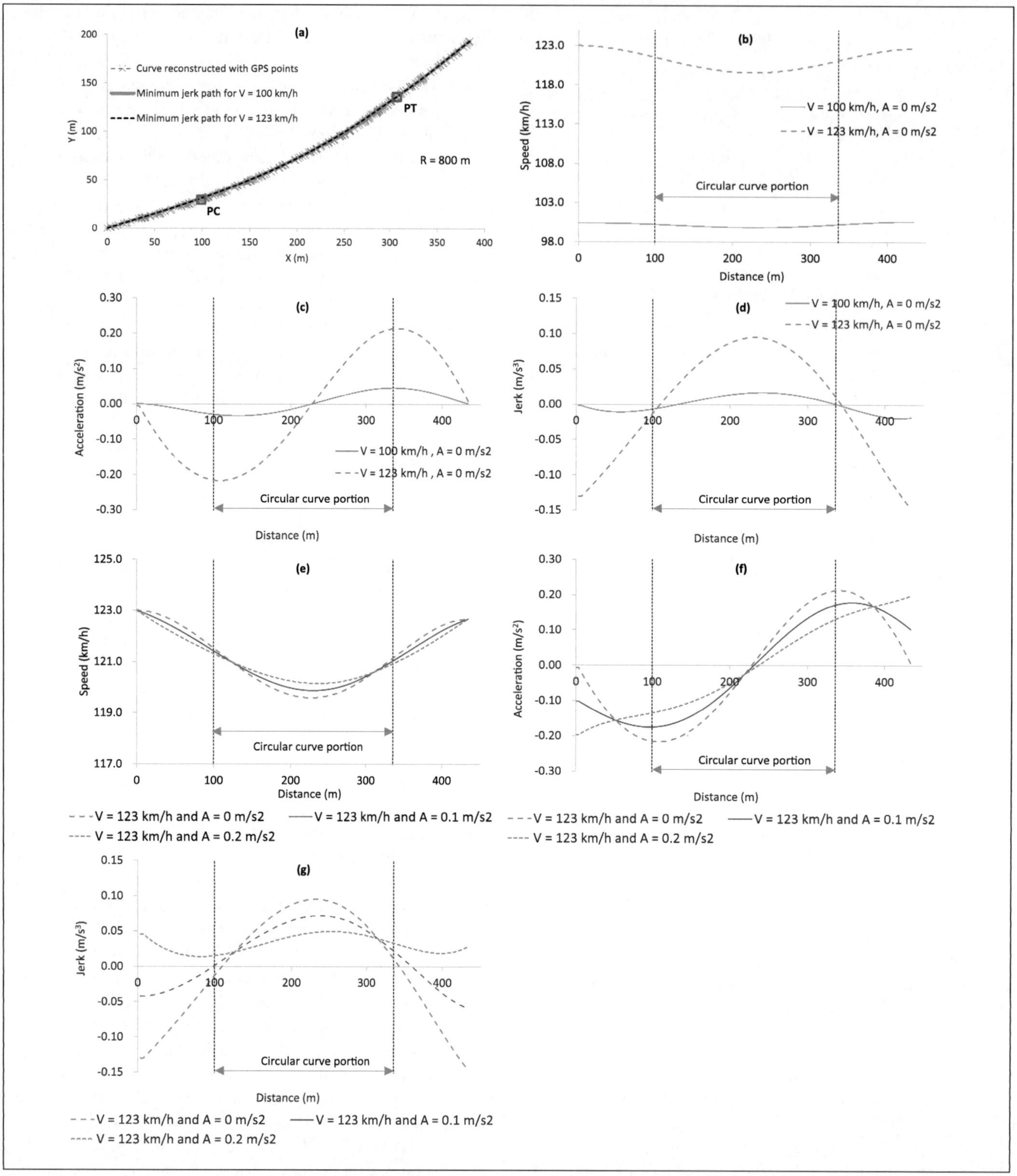

Figure 7. Comparison of (*a*) vehicle paths; (*b, e*) speed profiles; (*c, f*) acceleration profiles; (*d, g*) jerk profiles for different entry speed and acceleration levels for curve ID 5 (R = 800 m curve). *Note:* V = entry and exit speeds; A = entry and exit accelerations (or decelerations).

constraint or inputted into this modeling approach. Such possibilities could also be explored in future research.

Acknowledgments

The authors would like to thank the Central Nippon Expressway Company, Japan for providing data for this study.

Author Contributions

The authors confirm contribution to the paper as follows: study conception and design: Dias, Oguchi; data collection: Oguchi, Wimalasena; analysis and interpretation of results: Dias; draft manuscript preparation: Dias, Oguchi. All authors reviewed the results and approved the final version of the manuscript.

References

1. *A Policy on Geometric Design of Highways and Streets*, American Association of State Highway and Transportation Officials, Washington, D.C., 2001.
2. TRB Operational Effects of Geometrics Committee. *Modeling Operating Speed: Synthesis Report, Transportation Research Circular E-C151*, Transportation Research Board of the National Academies, Washington, D.C., 2011.
3. Wooldridge, M. D., K. Fitzpatrick, D. W. Harwood, I. B. Potts, L. Elefteriadou, and D. J. Torbic. *NCHRP Report 502: Geometric Design Consistency on High-Speed Rural Two-Lane Roadways*. Transportation Research Board of the National Academies, Washington, D.C., 2003.
4. Lamm, R., E. M. Choueiri, J. C. Hayward, and A. Paluri. Possible Design Procedure to Promote Design Consistency in Highway Geometric Design on Two-lane Rural Roads. *Transportation Research Record: Journal of the Transportation Research Board*, 1988. 1195: 111.
5. Lamm, R., and E. M. Choueiri. Recommendations for Evaluating Horizontal Design Consistency Based on Investigations in the State of New York. *Transportation Research Record: Journal of the Transportation Research Board*, 1987. 1122: 68–78.
6. Abdelwahab, W. M., M. T. Aboul-Ela, and J. F. Morrall. Geometric Design Consistency Based on Speed Change on Horizontal Curves. *Road and Transport Research*, Vol. 7, No. 1, 1998, pp. 13–23.
7. Fitzpatrick, K., L. Elefteriadou, D. W. Harwood, J. M. Collins, J. McFadden, I. B. Anderson, R. Krammes, N. Irizarry, K. D. Parma, K. M. Bauer, and K. Passetti. *Speed Prediction for Two-lane Rural Highways*. Report FHWA-RD-99-171. FHWA, U.S. Department of Transportation, Washington, D.C., 1999.
8. Morrall, J., and R. J. Talarico. Side Friction Demanded and Margins of Safety on Horizontal Curves. *Transportation Research Record: Journal of the Transportation Research Board*, 1994. 1435: 145–152.
9. Misaghi, P., and Y. Hassan, Y. Modeling Operating Speed and Speed Differential on two-lane Rural Roads. *Journal of Transportation Engineering*, Vol. 131, No. 6, 2005, pp. 408–418.
10. Zuriaga, A., A. García, F. Torregrosa, and P. D'Attoma. Modeling Operating Speed and Deceleration on Two-lane Rural Roads with Global Positioning System Data. *Transportation Research Record: Journal of the Transportation Research Board*, 2010. 2171: 11–20.
11. Cafiso, S., and G. Cerni. New Approach to Defining Continuous Speed Profile Models for Two-lane Rural Roads. *Transportation Research Record: Journal of the Transportation Research Board*, 2012. 2309: 157–167.
12. Montella, A., L. Pariota, F. Galante, L. Imbriani, and F. Mauriello. Prediction of Drivers' Speed Behavior on Rural Motorways Based on An Instrumented Vehicle Study. *Transportation Research Record: Journal of the Transportation Research Board*, 2014. 2434: 52–62.
13. Yang, L., and Y. Hassan. Driver Speed and Acceleration Behavior on Canadian Roads. Presented at 87th Annual Meeting of the Transportation Research Board, Washington, D.C., 2008.
14. Hu, W., and E. T. Donnell. Models of Acceleration and Deceleration Rates on a Complex Two-Lane Rural Highway: Results from a Nighttime Driving Experiment. *Transportation Research Part F: Traffic Psychology and Behaviour*, Vol. 13, No. 6, 2010, pp. 397–408.
15. Montella, A., F. Galante, L. L. Imbriani, F. Mauriello, and M. Pernetti. Simulator Evaluation of Drivers' Behaviour on Horizontal Curves of Two-lane Rural Highways. *Advances in Transportation Studies: An International Journal*, Vol. 34, 2014, pp. 91–104.
16. Montella, A., F. Galante, F. Mauriello, and M. Aria. Continuous Speed Profiles to Investigate Drivers' Behavior on Two-Lane Rural Highways. *Transportation Research Record: Journal of the Transportation Research Board*, 2015. 2521: 3–11.
17. Palmberg, A., J. Imberg, and R. Thomson. The Effect of Curve Geometry on Driver Behaviour in Curves by Using Naturalistic Driving Data. *Proc., 3rd International Symposium on Future Active Safety Technology towards Zero Traffic Accidents (FAST-zero 2015)*, Gothenburg, Sweden, 2015.
18. Bosetti, P., M. Da Lio, and A. Saroldi. On Curve Negotiation: From Driver Support to Automation. *IEEE Transactions on Intelligent Transportation Systems*, Vol. 16, No. 4, 2015, pp. 2082–2093.
19. Hicheur, H., S. Vieilledent, M. J. E. Richardson, T. Flash, and A. Berthoz. Velocity and Curvature in Human Locomotion Along Complex Curved Paths: A Comparison with Hand Movements. *Experimental Brain Research*, Vol. 162, No. 2, 2005, pp. 145–154.
20. Dias, C., M. Iryo-Asano, and T. Oguchi. Predicting Optimal Trajectory of Left-Turning Vehicle at Signalized Intersection. *Transportation Research Procedia*, Vol. 21, 2017, pp. 240–250.
21. Flash, T., and N. Hogan. The Coordination of Arm Movements: An Experimentally Confirmed Mathematical Model. *Journal of Neuroscience*, Vol. 5, No. 7, 1985, pp. 1688–1703.
22. Outline of ETC 2.0 - ETC portal site. *https://www.go-etc.jp/english/etc2/index.html*. Accessed May 22, 2017.
23. Wimalasena, K., C. Dias, T. Oguchi, K. Wada, and M. Iryo-Asano, Effect of Road Geometry on Free-Flow Speed: An Empirical Analysis using ETC 2.0 Data. *Proc., 53th Annual Meeting of Infrastructure Planning (JSCE)*, Hokkaido, Japan, 2016.
24. Pham, Q. C., H. Hicheur, G. Arechavaleta, J. P. Laumond, and A. Berthoz, A. The Formation of Trajectories During Goal-oriented Locomotion in Humans. II. A Maximum Smoothness Model. *European Journal of Neuroscience*, Vol. 26, No. 8, 2007, pp. 2391–2403.

25. Dias, C., M. Sarvi, and M. Iryo-Asano. Predicting Optimal Trajectories for Constrained Pedestrian Turning Manoeuvres. *Proc., 8th International Conference on Pedestrian and Evacuation Dynamics (PED2016)*, Hefei, China, 2016, 15-1-9.
26. Wada, Y., and M. Kawato. A Via-point Time Optimization Algorithm for Complex Sequential Trajectory Formation. *Neural Networks*, Vol. 17, No. 3, 2004, pp. 353–364.
27. Aloulou, A., and O. Boubaker, O. Minimum Jerk-based Control for a Three Dimensional Bipedal Robot. In *Intelligent Robotics and Applications. ICIRA 2011. Lecture Notes in Computer Science* (Jeschke, S., Liu, H., and Schilberg, D. eds.), Vol. 7102. Springer, Berlin, Heidelberg, 2011, pp. 251–262.
28. Bianco, L., C. Guarino, and M. Romano. Bounded Velocity Planning for Autonomous Vehicles. *Proc., IEEE/RSJ International Conference on Intelligent Robots and Systems (IROS 2005)*, IEEE, 2005, pp. 685–690.
29. Shamir, T. How Should an Autonomous Vehicle Overtake a Slower Moving Vehicle: Design and Analysis of an Optimal Trajectory. *IEEE Transactions on Automatic Control*. Vol 49, No. 4, 2004, pp. 607–610.
30. Hiraoka, T., T. Kunimatsu, O. Nishihara, and H. Kumamoto. Modeling of driver following behavior based on minimum-jerk theory. *Proc., 12th World Congress ITS*, San Francisco, USA, 2005.
31. Altamira, A., Y. García, T. Echaveguren, and J. Marcet. Acceleration and Deceleration Patterns on Horizontal Curves and Their Tangents on Two-lane Rural Roads. Presented at 93rd Annual Meeting of the Transportation Research Board, Washington, D.C., 2014.
32. Pérez-Zuriaga, A. M., F. J. Camacho-Torregrosa, and A. García. Tangent-to-curve Transition on Two-lane Rural Roads Based on Continuous Speed Profiles. *Journal of Transportation Engineering*, Vol. 139, No. 11, 2013, pp. 1048–1057.
33 Kim, S., and J. Choi. Effects of Preceding Geometric Conditions on Operating Speed Consistency of Multilane Highways. *Canadian Journal of Civil Engineering*, Vol. 40, No. 6, 2013, pp. 528–536.

The Standing Committee on Operational Effects of Geometrics (AHB65) peer-reviewed this paper (18-00775).

Article

TRR
JOURNAL OF THE TRANSPORTATION RESEARCH BOARD

Transportation Research Record
2018, Vol. 2672(17) 61–71

DOI: 10.1177/0361198118797186
journals.sagepub.com/home/trr
SAGE

Safety Evaluation for Turnarounds at Diamond Interchanges: Assessing the Texas U-Turn

Karen K. Dixon[1], Raul E. Avelar[1], Maryam Shirinzadeh Dastgiri[1], and Bahar Dadashova[1]

Abstract

Texas frontage road turnarounds at diamond interchange locations are a common treatment in the State of Texas. This configuration, also often referred to as the Texas U-turn, allows vehicles traveling on a one-way frontage road to access the opposing direction one-way frontage road via a U-turn before the terminal intersections at the diamond interchange. This technique removes two potential left-turn maneuvers from the adjacent at-grade intersections. The frontage road turnaround has operational benefits that result from shifting vehicles that would typically be occupying the intersection. The safety of these turnaround configurations, however, is not well known. This paper focuses on the observed safety performance of these turnaround configurations at diamond freeway interchanges in Texas. The authors first developed a large randomly sampled data set to facilitate a statistically reliable assessment of U-turn safety performance for Texas interchanges. Next they conducted a safety performance assessment at locations with and without turnarounds and determined that site features that significantly influence the number of crashes include the volume of cross street traffic, the cross street right-turn configuration at frontage roads, the minimum radius in the turnaround, the longitudinal distance from the U-turn exit to the closest downstream driveway, and the number of lanes on each frontage road. The safety performance findings at locations with and without turnarounds suggests that an agency could construct a turnaround to enhance facility operations with the knowledge that construction will not adversely impact safety. Although the research introduced in this paper focuses on Texas locations, the findings should be applicable to similar facilities in other states if the turnaround conditions are similar.

Frontage road turnarounds are located at major interchanges in Texas in urban and suburban regions (*1*). For a turnaround to be a feasible option, the frontage roads located parallel to the freeway lanes must have one-way operations. At locations where the freeway lanes are depressed, the turnaround is located on a structure adjacent to the arterial cross street lanes. When the freeway is elevated, the turnaround location will be under a bridge as shown in Figure 1. For both conditions, sight distance and visibility are important considerations that influence the U-turn geometry.

In 2010, the American Association of State Highway and Transportation Officials (AASHTO) published the first edition of the *Highway Safety Manual* (*2*). This document introduced the concept of integrating safety prediction based on statistically reliable safety performance functions that uniquely assess key characteristics of a roadway countermeasure or configuration. Unfortunately, safety research consistent with the techniques recommended in the *Highway Safety Manual* is not yet available for the Texas turnaround configuration. Consequently, transportation professionals do not have the ability to directly assess the number of expected crashes at an interchange location or determine how the configuration of the U-turn can influence the number of crashes. The research summarized in this paper, therefore, presents the results of a recent Texas turnaround safety assessment performed on behalf of the Texas Department of Transportation (TxDOT). Although the analysis focuses on Texas interchanges, the findings from this study are likely to be

[1]Texas A&M Transportation Institute, College Station, TX

Corresponding Author:
Address correspondence to Karen K. Dixon: k-dixon@tti.tamu.edu

Figure 1. Example of a Texas turnaround underpass configuration.
Source: Photo by Karen Dixon, Aerial View by GoogleEarth.

representative of similar interchange configurations in other states.

Database Development

Currently, the State of Texas does not maintain a comprehensive intersection or interchange database. To overcome this limitation, the authors first developed a study sample by identifying candidate study sites using geographic information system (GIS) merging techniques and then developing a geographically stratified random sample for database development and subsequent safety assessment.

Identifying Candidate Study Sites

The authors used existing roadway functional system information identified in the TxDOT roadway inventory database as an initial step toward determining freeway and arterial networks where diamond interchanges could be potentially located. Using GIS tools, the authors identified locations where interstate, freeway, or expressway facilities crossed arterial roadways. The authors developed separate GIS shape files based on these road categories and identified locations for potential points of the intersection that could represent interchanges or intersections. These potential intersections included intersecting points as far as 200 ft (approximately 60 m) from the freeway centerline. Using manual evaluations, the authors then confirmed which intersecting points did represent intersections between freeways and arterials. Figure 2 depicts these identified intersections for the selected roadway network (freeway and arterial).

Following this initial GIS screening for intersecting points, the authors next matched the large number of intersecting points (11,289 identified points) to potential locations where an interchange was likely. To narrow the search, an intersection with multiple points (generally four to six points) located close together often represented some sort of interchange. As can be observed in Figure 3*a*, an interchange with a U-turn will normally have at least six intersecting points. Therefore, to identify these potential interchange locations, the authors applied an additional 300 ft (approximately 100 m) buffer to all intersecting points as shown in Figure 3*b* to measure the distance between two intersecting points of randomly selected interchanges. As the objective of this activity was to include interchanges with and without U-turns, a diamond interchange with four or more intersection points within this buffer region represented a potential U-turn location. Finally, the authors filtered locations and assigned a single consolidated intersecting point to represent the latitude and longitude for the potential study interchange.

Developing a Stratified Random Sample

The selection of interchange locations that represent the larger diamond interchange population requires a random selection of the study locations. As the State of Texas is very large, the sampling procedure considered geographic location so that the resulting sample could be representative of the diamond interchanges across the state. The final TxDOT project report provides

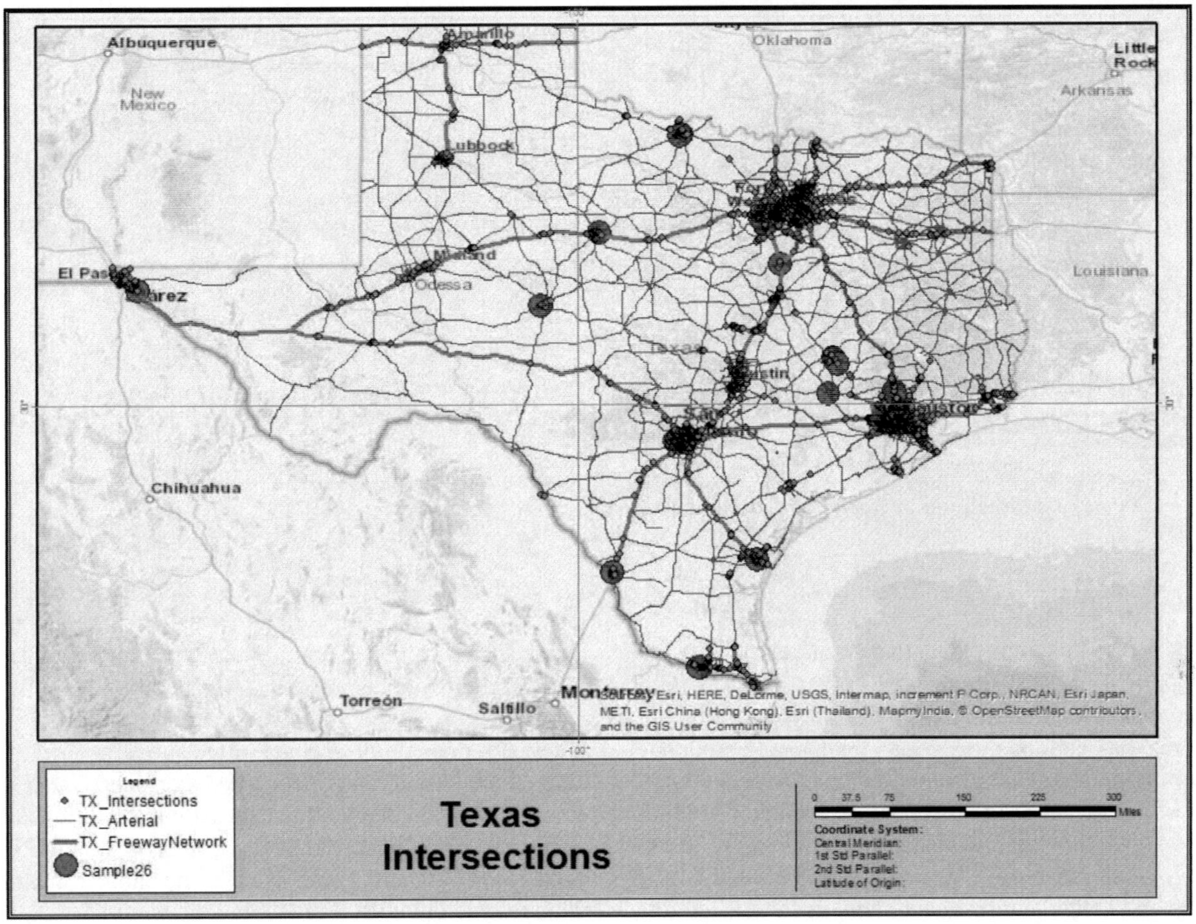

Figure 2. Locating an interchange with a U-turn (*a*) aerial view and (*b*) applying 300 ft buffer.

considerable detail about this sampling technique, but the final stratified random sample included 168 diamond interchanges with U-turns and 60 prospective diamond interchanges without U-turns. The authors used this data set for the subsequent site selection and data collection activities. The geographically stratified random sample included a two-stage process in which the authors first randomly selected interchanges from the entire state and then randomly selected additional interchanges so that the dataset included at least one interchange from every TxDOT district with freeway facilities. The initial interchange identification utilized geographic shapefiles and locations where arterials and freeway vectors crossed followed by photographic confirmation of each diamond interchange.

Data Collection and Database Development

Following the identification of the prospective study sites, the authors acquired data for each candidate site that included geometric data acquired from the TxDOT roadway inventory database and was supplemented or confirmed using aerials from Google Earth Pro®. In addition, the database included average annual daily traffic (AADT), the K-factor (for converting daily traffic volume proportions during peak hours), and the directional distribution factor known as the D-factor. This D-factor proved to be a valuable way to confirm one-way frontage road operations. In addition, the authors assigned crash data acquired from the Texas Crash Records Information System for the years 2009–2015.

The randomly selected study sites included a mixture of locations with and without U-turns. For locations with U-turns, the authors reviewed historic aerial photographs to confirm the presence of the U-turn for the entire seven-year period. For locations where the U-turn construction occurred during this study period, the authors removed any unconfirmed crash years. As an example, if an aerial photograph indicated a U-turn was not present in 2010 but was present in 2013, the database

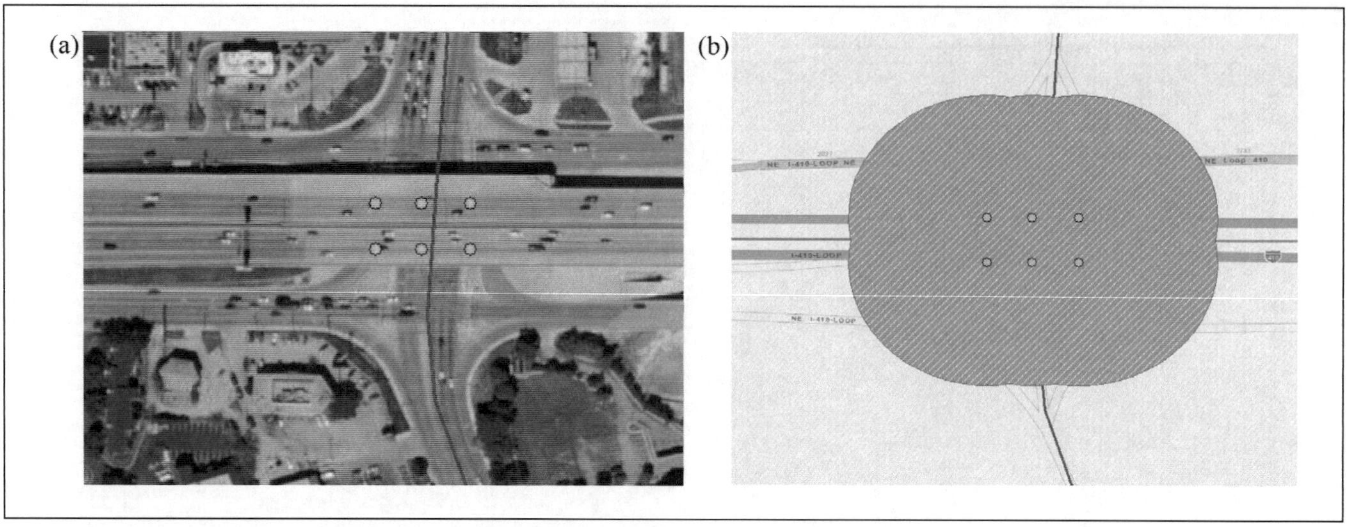

Figure 3. Locating an interchange with a U-turn (*a*) aerial view and (*b*) applying 300 ft buffer.

excluded crash data for 2011 and 2012. In addition, depending on the date of the aerial photograph, 2010 and 2013 crash data may have also been removed if the "before" site aerial was not in December or the "after" site aerial was not in January of the study year.

The analysis database included many potential site variables. Many of these did not prove to be significant and are not included for the purposes of brevity (including frontage road AADT, which the research team determined provided inaccurate values when compared to the crossroad AADT). Key data elements influential to the safety performance include the following data elements:

- Number of total lanes on each frontage road approach;
- Longitudinal distance between the U-turn exit and the first downstream driveway;
- Presence of traffic signal at the study intersections.
- Arterial right-turn treatment onto frontage roads.
- Depressed or elevated U-turn configuration; and
- U-turn leg dimensions (radii, lane widths, and length of dedicated U-turn).

In some cases, the data reduction process could not determine all site characteristics for a location. Often this constraint was a result of the interchange configuration. For example, U-turn leg dimensions could not always be determined at underpass turnaround locations because of occlusion from the structure. The authors elected to retain these sites and use a varying sample size during the statistical analysis stepwise variable assessment. The following sections review each of these roadway characteristics.

Number of Total Lanes on Each Frontage Road Approach. For each side of a study site (referred to from this point forward as a half site), the authors determined the number of frontage road main lanes at the intersection approach. Figure 4 depicts the total number of lanes as FR_A and FR_B. The authors used the labels "A" and "B" to identify each half site for documentation purposes. They do not imply specific directional information. It should be noted that most frontage roads had two to three lanes near the U-turn locations. It should also be noted that the study included 203 half sites with turnarounds and 133 half sites that did not have turnarounds. This resulted in a total of 336 half sites.

Longitudinal Distance Between the U-turn Exit and the First Downstream Driveway. The longitudinal distance to the first downstream driveway, measured from the point on the frontage road where the U-turning vehicles exit the U-turn and merge onto the frontage road, can influence the weaving and merging conditions on the frontage road. This label is represented as DWY_A and DWY_B in Figure 4. For this study, the authors measured this distance for the driveways that were within 500 ft of the U-turn exit. Out of 203 half sites, 53 of the locations had driveways positioned outside of this 500 ft threshold. The authors assigned a default DWY value of 500 ft for these locations.

Presence of a Traffic Signal at Study Intersection. The authors used aerial photographs and the companion StreetView tool to determine the type of traffic control present at each frontage road and cross street intersection location. At locations with a traffic signal, this method did not permit acquiring signal timing information. Approximately 93% of the intersections had traffic signals. Ultimately, the authors included only signalized intersection locations in the subsequent safety assessment.

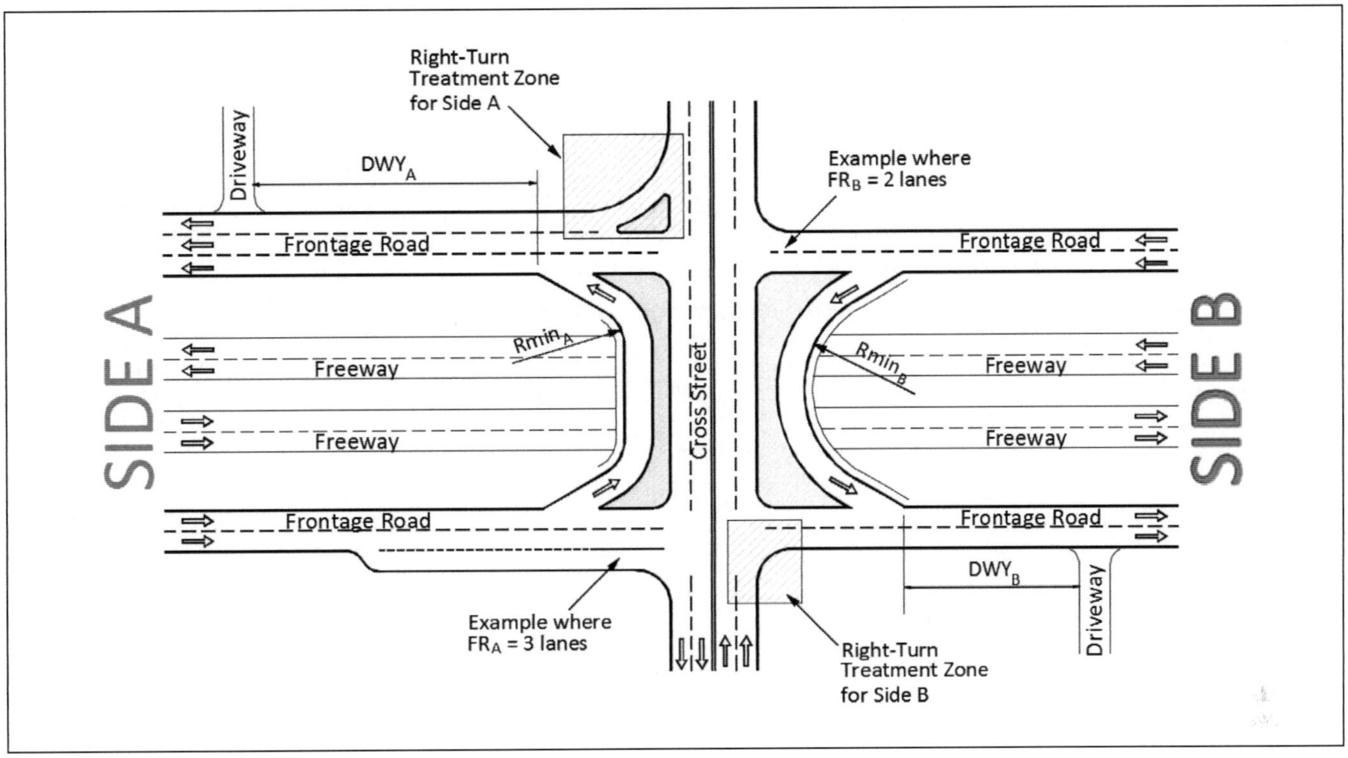

Figure 4. Turnaround configuration and influential site characteristics.

Arterial Right-Turn Treatment onto Frontage Roads. As a vehicle that is turning right from the cross street onto the one-way frontage road may potentially encounter a conflict with a vehicle exiting the U-turn, the authors categorized the configuration of the cross street right-turn treatment "zone" for each one-way frontage road location. Each configuration included the following two variables:

- Right-turn treatment "zone" entrance (see Figure 5), and
- Right-turn treatment "zone" exit (see Figure 6).

The entrance treatments included shared and exclusive cross street right-turn lanes with and without channelization islands. Ultimately, the authors tested the influence of these entrance treatments and determined that only Option A (a shared right with no island) was significantly associated with fewer crashes.

The exit treatment configurations required vehicles to merge into existing travel lanes or the geometric configuration included an additional lane (either as an acceleration lane or as another full lane). Even though most of the study locations had traffic signals, exit treatments at locations with islands often included supplemental traffic control (usually a yield or stop sign).

Depressed or Elevated Interchange Configuration. Of the 336 half site locations, 300 (150 of 168 sites) of them had arterial crossroads located below a freeway bridge, and 36 (18 of 168 sites) were elevated above the freeway. Although the placement as elevated versus depressed did not prove to be significant to intersection safety, the roadway geometry was directly influenced by these factors.

U-Turn Leg Dimensions (Widths and Lengths). Owing to the orientation of the U-turns, interchange bridges, and similar characteristics, the authors acquired, where feasible, the turning radius for the U-turn, the turning bay length, and the lane length. Although the authors evaluated the significance of all of these candidate variables, the minimum U-turn radius proved to be the only critical variable that influenced safety performance.

Analysis and Results

The authors performed a statistical analysis for the total crash condition (at turnaround locations) as well as the injury crash condition (denoted as the KAB model in which "K" represents a fatal crash, "A" represents an incapacitating injury crash, and "B" represents a non-incapacitating injury crash). The analytical approach used for this analysis was consistent with that used in a variety of similar studies (*3–5*). The following section first reviews the descriptive statistics for the larger dataset,

Table 1. Summary of Data Characteristics for a Site-Period

Variable name	Mean	SD	Min	Max	Total	N
Summary of observed crashes						
Total crashes	25.59	23.53	0	118	25,999	1016
Total crashes (known coordinates)	25.38	23.36	0	117	25,783	1016
KAB crashes	7.22	6.87	0	41	7,336	1016
Crashes involving a left-turning vehicle from frontage road	3.72	5.24	0	34	3,777	1016
Crashes involving a left-turning vehicle from frontage road (known coordinates)	3.69	5.23	0	34	3,754	1016
KAB crashes involving a left-turning vehicle from frontage road	0.89	1.47	0	12	908	1016
KAB crashes involving a left-turning vehicle from frontage road (known coordinates)	0.89	1.46	0	11	903	1016
Additional variable characteristics						
Minimum posted speed limit	45.4	5.9	30	55	46,145	1016
Maximum posted speed limit	46	6	30	55	46,705	1016

Note: Min = minimum; Max = maximum; SD = standard deviation; *N* = number.

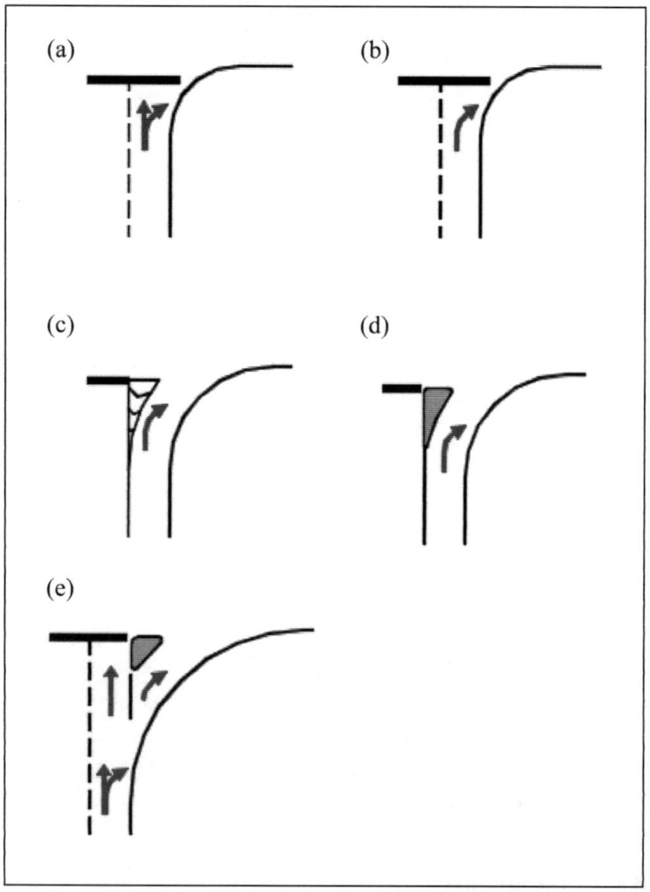

Figure 5. Right-turn treatment "zone" entrance options (*a*) shared right, no island, (*b*) exclusive right, no island, (*c*) exclusive right, painted island, (*d*) exclusive right, raised island, and (e) shared right, raised island, large radius.

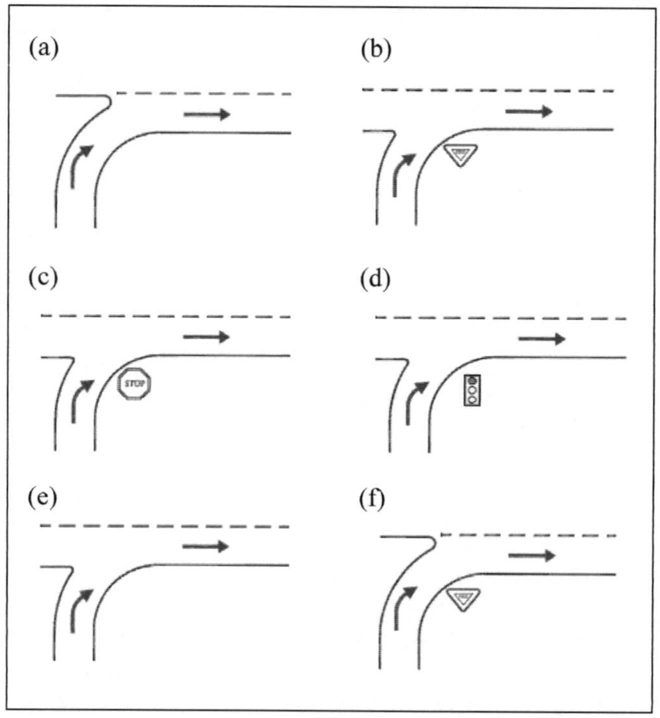

Figure 6. Right-turn treatment "zone" exit options (*a*) add lane, no additional control; (*b*) merge, yield control; (*c*) merge, stop control; (*d*) merge, signal control; (*e*) merge, no additional control; and (*f*) add lane, yield control.

then identifies some of the statistical modelling challenges addressed, and finally presents the resulting statistical models. Both models predicted crashes for the non-freeway crashes at the entire diamond interchange (rather than crashes on just one side).

Descriptive Statistics

Based on the premise that the number of left-turn crashes that originated on a frontage road can be expected to be reduced at U-turn locations, the authors inspected the crash data (for the years 2009–2015) and site data using a variety of target crash conditions including the proportion of crashes in which the vehicle originated on the frontage road. In some cases, missing data required slight modifications in sample sizes. This occurred when a data element could not be determined using the aerial photos and the roadway inventory file. For the initial inspection, the data included 2019 site-periods (seven potential years per site) from 164 sites; however, 77 of the site-periods did not have available speed limit data and so the authors could not evaluate these sites relative to speed limit. This resulted in 1016 site-periods from 152 sites with turnarounds (see Table 1).

Statistical Analysis

Following the initial data inspection, the authors performed a stepwise regression analysis to assess the influence of significant variables on the total number of crashes, as well as crash severity for the dedicated U-turn lane locations. During this process, the authors noted that only a small number of the interchanges had a posted speed limit of 30 mph. Most of the intersections had speed limits of 35 to 55 mph with an average posted speed limit of 45 to 46 mph. None of the 30 mph intersections had traffic signals for both of the intersections associated with the interchange. To mitigate the influence of this subset of study sites, the authors elected to remove these intersections from the subsequent analysis.

Model Development. The crash model development focused on the identification of statistically significant variables and an assessment of the best functional form that represents each identified variable. For example, the use of logarithmic adjustments may be appropriate for some variables that do not follow a linear format when graphically plotted. The authors used generalized mixed-effects models to compensate for more than one study year of data at a specific site location. The authors also explored an extensive number of variables during the variable selection. Of particular note is the selected traffic volume variable. It is feasible to consider freeway volume, cross street volume, frontage road volume (for both frontage roads), and U-turn volume. Ultimately, the cross street AADT values proved to be the most influential traffic volume variable for this assessment. The authors used a negative binomial model approach for this analysis.

Total Crash Model. The authors used a "sample expansion" step before switching subsets of potential independent variables as the initial model development approach (*6–8*). For the development of the final predictive model, the authors focused on signalized intersection locations with speed limits greater than 30 mph. Through the application of stepwise regression procedures, the authors then developed a final predictive model for total crashes. The model selection occurred in several stages so that the authors could consider groups of variables jointly available for data subsets. Once a modelling stage had arrived at the point with the best fit and the fewest variables (referred to as a parsimonious model), the authors fit the model to the largest subset of data that included all the variables in the model specification. This iterative process continued until the authors had considered all variables at least two times for inclusion into the model. The resulting dataset represented 86 site locations with 459 site-periods available for model estimation. Table 2 shows the final total crash model. The authors applied scaling adjustments to two of the variables with more widely dispersed data (*9–11*). The cross street right-turn entrance treatment A (shared right with no island) proved to be more significant that the other entrance treatments. The cross street right-turn exit treatments were not statistically significant for the total crash model.

A common goodness of fit assessment gives the cumulative residual (CURE) plot. Optimally, the CURE plot for each variable should oscillate around the line that represents zero. For the total crash model, these plots depict minimal deviations beyond the expected boundaries for key variables (see Figure 7). As can be observed, the expected oscillation occurs for the lower volume thresholds, but the cumulative residuals appear to be skewed for the higher volume locations.

The final model presents a functional form that incorporates scaling of some variables. This total crash model is

$$N_{Total} = e^{\left[5.304-(0.271\times RtA)+(0.703\times AvgLn)-\left(0.268\times\frac{DWY-196.319}{155.752}\right)+0.113\times\left[\frac{CS_{AADT}-13,516.82}{10,059.57}\right]-(0.951\times\ln(Rmin))\right]}$$

The equation included a scaling adjustment for the DWY and the CS_{AADT} variables. The authors adjusted these scaled variables by subtracting the mean value and then dividing by the standard deviation of the variable. The reduced N_{Total} model is:

$$N_{Total} = \frac{e^{\left[5.491-(0.271\times RtA)+(0.703\times AvgLn)-(0.0017\times DWY)+\left(1.124\times 10^{-5}\right)\times CS_{AADT}\right]}}{Rmin^{0.9512}}$$

Table 2. Predictive Model for Total Crashes (Signalized Sites)

Continuous variable descriptive statistics

Variable name	Description	Mean	SD	Min.	Max.	Total	N
AvgLn	Average number of frontage road lanes FR_A and FR_B (see Figure 4)	2.6	0.5	2	4	1195	459
DWY	Distance to closest downstream driveway (ft)	196.3	155.8	10	500	90,110	459
CS_AADT	Cross street AADT (vpd)	13,516.8	10,059.6	200	54,609	6,204,220	459
Rmin	Minimum turning radius in U-turn (ft)	49.9	14.6	22	129	22,918	459

Total crash model

Variables	Estimate	Standard error	Z value	Pr (>|z|)	Significance[b]
(Intercept)[a]	5.3041	1.0862	4.8834	1.0428×10^{-6}	***
RtA	−0.2708	0.1023	−2.6480	0.0081	**
AvgLn	0.7027	0.1616	4.3490	0.0000	***
Scale (D_to_closest_driveway)	−0.2684	0.0719	−3.7320	0.0002	***
Scale (CS_AADT)	0.1131	0.0489	2.3120	0.0208	*
ln (Rmin)	−0.9512	0.2454	−3.8760	0.0001	***

Note: Min = minimum; Max = maximum; SD = standard deviation; *N* = number; RtA = Number of instances at the site where RtA had a shared right turn lane and no channelization island (see Figure 5). Value of RtA ranges from zero (no shared lane option) to two (shared lane option at both cross street right-turn locations).

[a]Includes adjustment random effects.

[b]Significance levels are as follows:

*Statistically different from 0.0 at the 5.0% significance level.

**Statistically different from 0.0 at the 1.0% significance level.

***Statistically different from 0.0 at the 0.1% significance level.

One interesting observation about the final (and ultimately the KAB) model is that the presence of a turnaround does not appear as a critical variable in the model. The authors included this variable in the stepwise analysis and it was not significant. This finding suggests that constructing a turnaround does not significantly affect the total number of non-freeway interchange crashes, and so this treatment should complement operational benefits of adding turnarounds at diamond interchange locations.

By inspection of the variables included in the total crash model for interchanges with turnarounds, the following general observations merit consideration:

- Locations where the right-turn from the cross street originates from a shared lane and does not have a large turning radius or a raised island (Option A) could result in 23.7% fewer crashes (calculated as $1-e^{(-0.2708)} = 0.237$). This finding is significant at 1.0%.
- For frontage roads with 2–4 lanes, the number of crashes increases by a factor of 2.01 (doubles) for each additional frontage road lane (calculated as $e^{(0.7027)} = 2.014$). This finding is significant at 0.1%. This finding is likely a surrogate for the varying frontage road AADT values.
- The number of crashes reduces as the distance to the closest downstream driveway increases. This reduction is approximately 1.7% for each additional 10 feet between the closest U-turn exit and the downstream driveway (calculated as $1 - e^{\frac{-0.2684}{155.752} \times 10} = 0.0171$). This finding is significant at 0.1%.
- The number of crashes increases by 1.1% for each additional 1000 vehicles per day (vpd) increase in cross street AADT (calculated as $1 - e^{\frac{0.1131}{10,059.6} \times 1000} = 0.011$). This finding is significant at 5.0%.
- The number of crashes decreases by 8.7% for each increase in 10% in the turning radius of the

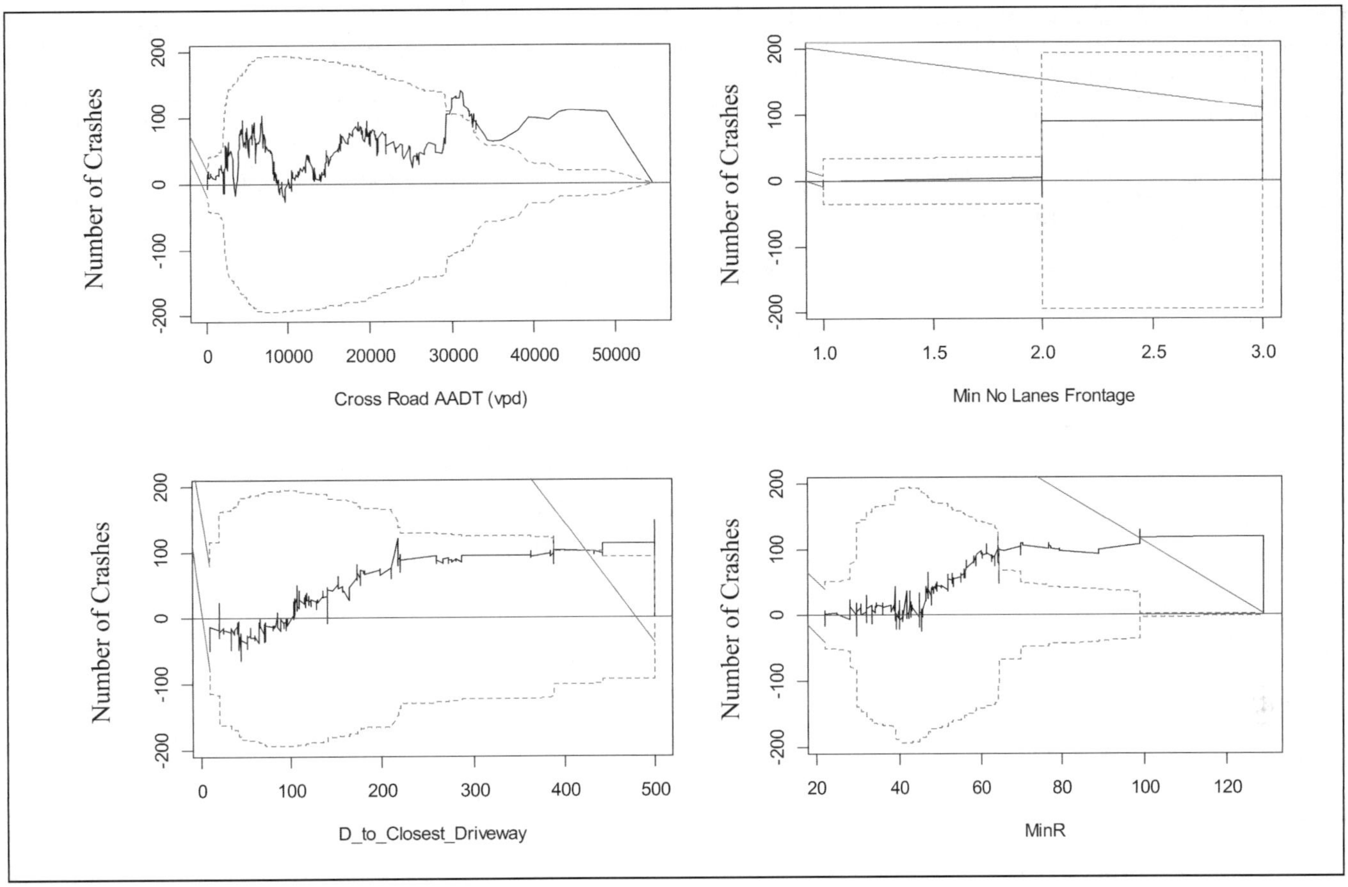

Figure 7. CURE Plots for the total crash model.

U-Turn (calculated as $1 - e^{-0.9512 \times \ln(1.1)} = 0.0867$). This finding is significant at 0.1%.

KAB Crash Model. The authors used a similar modelling approach as that applied for total crashes to evaluate the injury crashes at the Texas turnaround locations (see Table 3). The CURE plots for this model exhibited similar oscillating patterns to those as shown for total crashes and therefore are not shown for this model.

The KAB crash model equation is:

$$N_{KAB} = e^{[3.08-(0.31 \times RtA) + (0.48 \times AvgLn)-(0.18 \times Merge_{RT})-\left(0.20 \times \frac{DWY-191.58}{151.86}\right) + (0.072 \times \ln(CS\_AADT))-(0.67 \times \ln(Rmin))]}$$

After simplification, the resulting predictive KAB equation takes the following form:

$$N_{KAB} = \frac{CS\_AADT^{0.0719}}{Rmin^{0.6684}} \times e^{[3.32-(0.31 \times RtA) + (0.48 \times AvgLn)-(0.18 \times Merge_{RT})-(0.0013 \times DWY)]}$$

By inspection of the variables included in the KAB model for signalized interchanges with turnarounds, the following observations merit consideration:

- Sites where the right-turn from the cross street must share a lane have 26.4% fewer KAB crashes (calculated as $1-e^{(-0.3063)}$). This finding is significant at 1.0%.
- Sites where the right-turn traffic must merge with the frontage road traffic (without adding a lane) have fewer severe crashes (significant at < 0.01%). This merge configuration is associated with a reduction of 16.4% (calculated as $1-e^{(-0.1801)}$). This finding is significant at 1.0%.
- The number of KAB crashes is smaller by 1.3% for each additional 10 feet between the closest downstream driveway and the U-turn exit (calculated as $1 - e^{\frac{-0.1969}{151.86} \times 10} = 0.0126$). This finding is significant at 1.0%.

Table 3. Predictive Models for KAB Crashes (Signalized Intersections)

Continuous variable descriptive statistics for the KAB model

Variable name	Description	Mean	SD	Min.	Max.	Total	N
CS_AADT	Cross street AADT (vpd)	13,872.4	10,442.3	200	54,309	6,478,395	467
AvgLn	Average number of frontage road lanes FR_A and FR_B (see Figure 4)	2.62	0.45	2	4	1225	467
DWY	Distance to closest downstream driveway (ft)	191.54	151.86	10	500	89,467	467
Rmin	Minimum turning radius in U-turn (ft)	49.99	14.84	22	129	23,344	467

Final KAB model

Variables	Estimate	Standard error	Z value	Pr (>|z|)	Significance[b]
(Intercept)[a]	3.0758	1.1096	2.7720	5.5718E-03	.
ln (CS_AADT)	0.0719	0.0392	1.8360	6.6290E-02	.
RtA	−0.3063	0.1013	−3.0220	2.5100E-03	**
AvgLn	0.4797	0.1580	3.0350	2.4000E-03	**
$Merge_{RT}$	−0.1801	0.0892	−2.0200	4.3350E-02	*
Scale (DWY)	−0.1969	0.0699	−2.8190	4.8200E-03	**
ln (Rmin)	−0.6684	0.2412	−2.7710	5.6000E-03	**

Note: Min = minimum; Max = maximum; SD = standard deviation; *N* = number; RtA = Number of instances at the site where right-turn "zone" entrance treatment had a shared right turn lane and no channelization island (see Figure 5). Value of RtA ranges from zero (no shared lane option) to two (shared lane option at both cross street right-turn locations). $Merge_{RT}$ = Number of instances at the site where the right-turn "zone" exit treatment merged into an existing lane (see Figure 6). Value of $Merge_{RT}$ ranges from zero (only included added lanes) to two (all cross street right-turn lanes require vehicles to merge into an existing lane).

[a]Includes adjustment random effects.

[b]Significance levels are as follows:

*Statistically different from 0.0 at the 5.0% significance level.

**Statistically different from 0.0 at the 1.0% significance level.

***Statistically different from 0.0 at the 0.1% significance level.

Table 4. Influence of Site or Traffic Characteristics on Crashes

Site or traffic characteristic	Significant influence: Total crashes	Significant influence: KAB crashes
As the cross street AADT increases:	Total crashes increase	KAB crashes increase
Cross street right-turn maneuvers onto the frontage road that originate in a shared lane are associate with:	Fewer total crashes	Fewer KAB crashes
Cross street right-turning vehicles that merge into existing lanes are associated with:	No significantly noticeable change in total crashes	Fewer KAB crashes
As the number of lanes increases for each frontage road approach:	Total crashes increase	KAB crashes increase
As the longitudinal distance between the U-turn exit to the nearest downstream driveway increases:	Total crashes decrease	KAB crashes decrease
As the minimum U-turn radius increases:	Total crashes decrease	KAB crashes decrease

- The number of KAB crashes is smaller by 6.2% for each increase of 10% in the turning radius of the U-Turn (calculated as $1 - e^{-0.6684 \times \ln(1.1)} = 0.62$). This finding is significant at 1.0%.

Conclusions and Recommendations

This study examined the safety effects of turnarounds at diamond interchanges in Texas. Based on a statistical evaluation of total crashes and injury crashes, the authors generally concluded that key variables associated with crash frequency include cross street AADT, cross street right turn configuration, number of frontage road lanes, longitudinal distance from U-turn exit to nearest downstream driveway, and U-turn minimum radius values collectively influence the number of crashes at turnaround locations. The authors limited the study sites to locations with posted speed limits on the frontage road of 35 to 55 mph and with signalized intersection configurations. Table 4 summarizes how each of these site or traffic characteristics influences the number of predicted crashes.

In addition, the authors evaluated sites with and without turnarounds. During the statistical analysis, it became clear that the variable that indicated the presence of a turnaround was not statistically significant. This finding suggests that the construction of turnarounds at diamond interchanges will not substantially affect the total number of crashes at these locations. This means that the construction of a turnaround as a mechanism for improving operations and removing the two left-turns from adjacent signalized intersections will not have any significant adverse safety implications. In Texas, the construction of turnarounds is routine for newly constructed or retrofitted interchanges in urban areas. The findings of this study demonstrate that the construction of turnarounds for operational purposes will not introduce additional safety concerns.

Acknowledgments

The authors would like to thank the Texas Department of Transportation for their support of this research effort.

Author Contributions

The authors confirm contribution to the paper as follows: study conception and design: K. Dixon, R. Avelar; data collection: K. Dixon, M. Shirinzadeh Dastgiri, B. Dadashova; analysis and interpretation of results: K. Dixon, R. Avelar, M. Shirinzadeh Dastgiri, B. Dadashova; draft manuscript preparation: K. Dixon, R. Avelar. All authors reviewed the results and approved the final version of the manuscript.

References

1. TxDOT. *Roadway Design Manual*, Revised October 2014. http://onlinemanuals.txdot.gov/txdotmanuals/rdw/rdw.pdf. Accessed February 20, 2017.
2. American Association of State Highway Transportation Officials. *Highway Safety Manual*. American Association of State Highway Transportation Officials, Washington, D.C., 2010.
3. Bauer, K. M., D. W. Harwood, and K. R. Richard. *Safety Effects of Using Narrow Lanes and Shoulder-Use Lanes to Increase the Capacity of Urban Freeways*. Report No. FHWA-HRT-05-001. Federal Highway Administration, McLean, VA, 2005.
4. Bauer, K. M., D. W. Harwood, W. E. Hughes, and K. R. Richard. 2004. Safety Effects of Using Narrow Lanes and Shoulder-Use Lanes to Increase the Capacity of Urban Freeways. *Transportation Research Record: Journal of the Transportation Research Board*, 2004. 1897: 71–80.
5. Bonneson, J. A., S. Geedipally, M. Pratt, and D. Lord. *NCHRP Project No. 17-45: Safety Prediction Methodology and Analysis Tool for Freeways and Interchanges*. Transportation Research Board of the National Academies, Washington, D.C., 2012.
6. Fox, J., and S. Weisberg. *An {R} Companion to Applied Regression*, 2nd ed. SAGE Publications, Thousand Oaks, CA, 2011.
7. Hauer, E., and J. Bamfo. Two Tools for Finding what Function Links the Dependent Variable to the Explanatory Variables. *Proc., ICTCT 1997 Conference*. Lund, Sweden, 1997.
8. Hilbe, J. M. *Negative Binomial Regression*, 2nd ed. Cambridge University Press, Cambridge, United Kingdom, 2011.
9. Park, E., J. Park, and T. Lomax. A Fully Bayesian Multivariate Approach to Before-After Safety Evaluation. *Accident Analysis and Prevention*, Vol. 42, No. 4, 2010, pp. 1118–1127.
10. Persaud, B., B. Lan, C. Lyon, and R. Bhim. Comparison of Empirical Bayes and Full Bayes Approaches for Before-After Road Safety Evaluations. *Accident Analysis and Prevention*, Vol. 42, No. 1, 2010, pp. 38–43.
11. Ramsey, F. L., and D. W. Schafer. *The Statistical Sleuth: A Course in Methods of Data Analysis*, 2nd ed. Duxbury, Pacific Grove, CA, 2002.

The Standing Committee on Operational Effects of Geometrics (AHB65) peer-reviewed this paper (18-04662).

Article

TRR
JOURNAL OF THE TRANSPORTATION RESEARCH BOARD

Transportation Research Record
2018, Vol. 2672(17) 72–82

DOI: 10.1177/0361198118776111
journals.sagepub.com/home/trr
SAGE

Speed Prediction Models for Trucks on Horizontal Curves of Two-Lane Rural Roads

David Llopis-Castelló[1], Brayan González-Hernández[1], Ana María Pérez-Zuriaga[1], and Alfredo García[1]

Abstract

Road safety is closely related to geometric design consistency, which is usually assessed by examining operating speed. Most consistency models only consider passenger car speeds, even though the interaction between passenger cars and heavy vehicles plays a pivotal role in road safety. This is due to the fact that there are too few models to estimate heavy vehicle speeds. This study aims to develop speed prediction models for heavy vehicles on horizontal curves of two-lane rural roads. To do this, continuous speed profiles were collected by using Global Positioning System (GPS) tracking devices on 11 road sections. Truck speeds were analyzed on 105 horizontal curves. The results showed that the radius of the horizontal curve and the grade at the point of curvature have a significant influence on heavy vehicle speeds. In this regard, vertical alignment only has a significant effect on truck speeds along upgrades. In addition, different trends were identified for loaded and unloaded trucks, so different speed models were calibrated for each of them. As a result, heavy vehicle speeds were adversely affected by grades greater than 3%. This phenomenon was larger for loaded trucks than for unloaded ones. Finally, the calibrated 85th and 15th percentile speed models were compared with those developed previously. As a conclusion, the use of the proposed models in this study was recommended on Spanish two-lane rural roads due mainly to the different characteristics of heavy vehicles around the world.

Road safety is one of the major concerns in our society. In fact, around 1.2 million people die and 50 million are injured in road crashes every year (*1*).

Road crashes are mainly caused by three concurrent factors: infrastructure, vehicle, and human factors. Specifically, the interaction between the infrastructure and human factors has been thoroughly studied in recent years through geometric design consistency, which can be defined as how drivers' expectations and road behavior relate. In this way, a consistent road aims to minimize unexpected events to road users while driving along it, whereas an inconsistent road might present numerous surprises to drivers, leading to anomalous behavior and increasing the likelihood of crashes occurring.

The most commonly used methods to assess geometric design consistency are based on operating speed (V_{85}). This speed is defined as the 85th percentile of the distribution of speeds selected by drivers under free-flow conditions with no environmental restrictions. This speed can be estimated by means of operating speed models. Although there are many models which allow the estimation of the operating speed for passenger cars on two-lane rural roads, there are relatively few models for predicting the operating speed for heavy vehicles. For this reason, most of the consistency models are only based on the operating speed profile of passenger cars.

In addition, heavy vehicles should be better integrated into highway design guidelines. Although the Green Book produced by the American Association of State Highway and Transportation Officials (AASHTO) (*2*) takes them into account for pavement design, climbing lanes, superelevation rate, and speed deceleration on upgrades and downgrades, Spanish highway design guidelines only consider them for climbing lanes and pavement design. However, the interaction between both passenger cars and heavy vehicles plays a pivotal role in road safety, mainly on two-lane rural roads with a large percentage of heavy vehicles (*3*). Therefore, not

[1]Highway Engineering Research Group (HERG), Universitat Politècnica de València, Valencia, Spain

Corresponding Author:
Address correspondence to David Llopis-Castelló: dallocas@doctor.upv.es

considering heavy vehicle speeds might lead to inconsistent geometric designs.

Some authors have studied this phenomenon through the speed difference between passenger cars and heavy vehicles. According to Harwood et al. (*4*), this speed difference might cause inconsistencies in vehicle operation along upgrades. Likewise, Leisch and Leisch (*5*) suggested that this speed difference should be limited to 15 km/h.

On the other hand, several research studies have been focused on the calibration of operating speed models for heavy vehicles. Two types of model can be distinguished: those based on speed data collected on rural roads and those based on dynamic and cinematic performance.

Regarding the first model, Misaghi and Hassan (*6*) analyzed the influence of several geometric variables of horizontal curves, such as the radius or the longitudinal grade, on heavy vehicle speeds. Speed data were collected using electronic counter-classifiers on 20 horizontal curves of two-lane rural roads. As a result, two speed models were proposed to estimate the 85th percentile speed at midcurve with the horizontal curve radius as an explanatory variable. In addition, statistically significant differences between passenger car speeds and heavy truck speeds were identified, but not between passenger car speeds and light truck speeds.

Jacob and Anjaneyulu (*7*) studied the operating speeds of different types of vehicle (passenger cars, motorcycles, buses, and trucks) on 152 horizontal curves with grades between −2% and + 2% and significant differences among the speeds and speed reductions developed by the different types of vehicle were found. Separate models were, therefore, developed to analyze the influence of geometry on the speeds developed by the different types of vehicle. The results indicated that the radius and length of the horizontal curve have a significant influence on operating speeds. The greater the radius or length of the horizontal curve, the greater the speed differences between the different types of vehicle.

However, most researchers concluded that, while passenger car speed profiles can mainly be based on horizontal alignment, heavy vehicle speed profiles depend on both horizontal and vertical alignment. Therefore, the analysis of the operating speed differences between passenger cars and heavy vehicles should be carried out considering three-dimensional (3D) geometric effects. In this regard, Leisch and Leisch (*5*) developed a speed profile by combining the speed profile estimated by considering only the horizontal alignment (assuming level grade) and the speed profile only considering the vertical alignment (assuming no restrictions on the horizontal alignment).

Likewise, several regression models were developed by Donnell et al. (*3*) to predict the 85th percentile speed of heavy vehicles on horizontal curves. These models were calibrated using a combination of field data collected on 11 horizontal curves of two-lane rural roads and simulated data by means of TWOPAS. In this way, it was found that an increase in the horizontal curve radius and the approach tangent length was associated with greater operating speeds at the point of tangency. In addition, a high grade of the approach tangent was associated with a lower operating speed at this location.

Another study based on the horizontal and vertical alignments was developed by Gibreel et al. (*8*). In this case, speed data were collected for all types of vehicle under free-flow conditions on nine horizontal curves combined with a sag curve and 10 horizontal curves combined with a crest curve. The results showed that several parameters significantly affect the operating speed: radius and deflection angle of the horizontal curve, horizontal distance between the point of horizontal intersection and the point of vertical intersection, length of the vertical curve, grade, and superelevation rate.

On the other hand, Saifizul et al. (*9*) studied the influence of the size and weight of heavy vehicles on their speeds. In this regard, important speed differences were found when vehicles were significantly different in size. In addition, when vehicles were almost similar in size and only differed in the number of axles, the Gross Vehicle Weight (GVW) was the most influential factor. If GVW was lower than 20 t, the speed decreased as the weight decreased. By contrast, the speed remained without variation when GVW was greater than 20 t.

Other heavy vehicle speed profile models have been developed on upgrades based on dynamic and cinematic performance (*10–13*). These models were mainly used for the study of climbing lanes for heavy vehicles. These speed models depend on the grade, vehicle weight-to-power ratio (WPR), resistance of the air, pavement characteristics, rolling resistance coefficients, and friction coefficients.

Summarizing, few operating speed models for heavy vehicles have been calibrated. Most of these studies used speed spot data collected on a low number of horizontal curves for calibration and presented a large variation in model form, explanatory variables, and regression coefficients. This might be due to differences in driver behavior, mechanical characteristics of the vehicles, and road geometry. Thus, a single model is not universally accepted.

The present research analyzes truck speeds on horizontal curves of two-lane rural roads based on continuous speed profiles collected using Global Positioning System (GPS) tracking devices. The main advantage of this method is the large amount of continuous speed data that can be collected without significant influence on drivers.

Objective and Hypotheses

The objective of this study is to analyze the influence of road geometric design on heavy vehicle speeds and develop speed prediction models on horizontal curves of two-lane rural roads. This research not only focuses on the evaluation of the operating speed, but also analyzes the 15th percentile speed, since the low percentiles encourage the emergence of traffic conflicts between passenger cars and heavy vehicles.

The research is based on two hypotheses. The first one is that road grade, which has no influence on passenger car speeds, produces lower heavy vehicle speeds on upgrades. The second hypothesis is that the speed difference between heavy vehicles of similar weight is very low. Despite passenger car speeds being mainly influenced by the driver, heavy vehicle speeds are significantly influenced by the heavy vehicle's engine performance or its WPR.

Methodology

This research was based on continuous speed data collected through Global Positioning System (GPS) tracking devices. First, GPS devices were lent to three transport companies, which equipped their heavy vehicles with these devices. Next, the transport companies gave the devices back and the speed data were filtered and processed. In addition, information about the use of each GPS (day, time, and type of heavy vehicle) was also reported by the transport companies. The geometry of the two-lane rural roads was recreated by means of the methodology proposed by Camacho-Torregrosa et al. (*14*). Finally, different regression models were calibrated based on the geometric characteristics of the horizontal curves to estimate heavy vehicle speeds.

Data Collection

Speed data were collected between May 2015 and July 2015 in working days under dry weather conditions. Heavy vehicles from three transport companies were equipped with 1Hz pocket-sized GPS tracking devices. The accuracy of these GPS devices was 2.5 m. However, this accuracy is mainly composed of a general bias (common to all measurements) and a minimum random error. In fact, the bias can be addressed through moving all trajectories (it does not affect accuracy), while the second error is minimal (millimeters), thus being very easy to address.

Heavy vehicles performed round trips, loaded in one direction and unloaded in the other. All heavy vehicles were 5-axle single trailers. The WPR ranged between 35 and 54 kg/kW for unloaded trucks, with the average being 43 kg/kW. However, the average WPR increased to 120 kg/kW for loaded trucks.

Eleven two-lane rural road sections were identified from the path of the heavy vehicles equipped with GPS (Table 1). All road sections were located in the Valencian Region (Spain). The lane width ranged from 3.0 m to 3.5 m, while the shoulder width varied from 1.0 m to 1.5 m. The horizontal alignment of these two-lane rural road sections was recreated using an algorithm based on the heading direction (*14*). Additionally, the vertical alignment was obtained by means of the geometric road design software Civil 3D using LIDAR data provided by the National Plan for Aerial Orthography (PNOA) with a root mean square error of 20 cm in height (*15*).

The GPS devices stored the following data at a 1 Hz frequency: latitude, longitude, altitude, heading direction, time, and date.

Collected data were transformed to Universal Transverse Mercator (UTM) coordinates and filtered and processed to obtain individual continuous speed profiles (*16*). After that, free-flow conditions were checked (*17*). The test used is based on the hypothesis that every single driver behaves in a particular way, approaching their individual speed profile to a certain speed percentile. Therefore, for each individual speed profile, non-free-flow road sections were associated with sudden changes in its speed profile compared with its usual speed percentile profile.

After removing non-free-flow sections for every single driver, 85th and 15th percentile speed profiles were

Table 1. Parameters of the Road Sections

Road	Segmento name	Length (m)	Grade range (%)
CV-425	Buñol – Alborache	1,956	[−8; + 6]
CV-425	Alborache – CV-429	752	[−3; + 6]
CV-425	Macastre I – Macastre II	1,419	[−5; + 2]
CV-425	Macastre II – CV-580	11,996	[−9; + 11]
CV-425	La Matrona I – La Matrona II	5,836	[−12; + 12]
CV-345	Villar del Arzobispo – Higueruelas	7,215	[−5; + 8]
CV-600	Xávita – Fenollet	2,685	[−2; + 2]
CV-610	Genovés – Cuatretonda	7,304	[−8; + 10]
CV-610	Cuatretonda – Llutxent	2,686	[−3; + 10]
CV-608	Llutxent – Planta	1,660	[−8; + 8]
CV-610	Llutxent – CV-60	5,685	[−5; + 6]

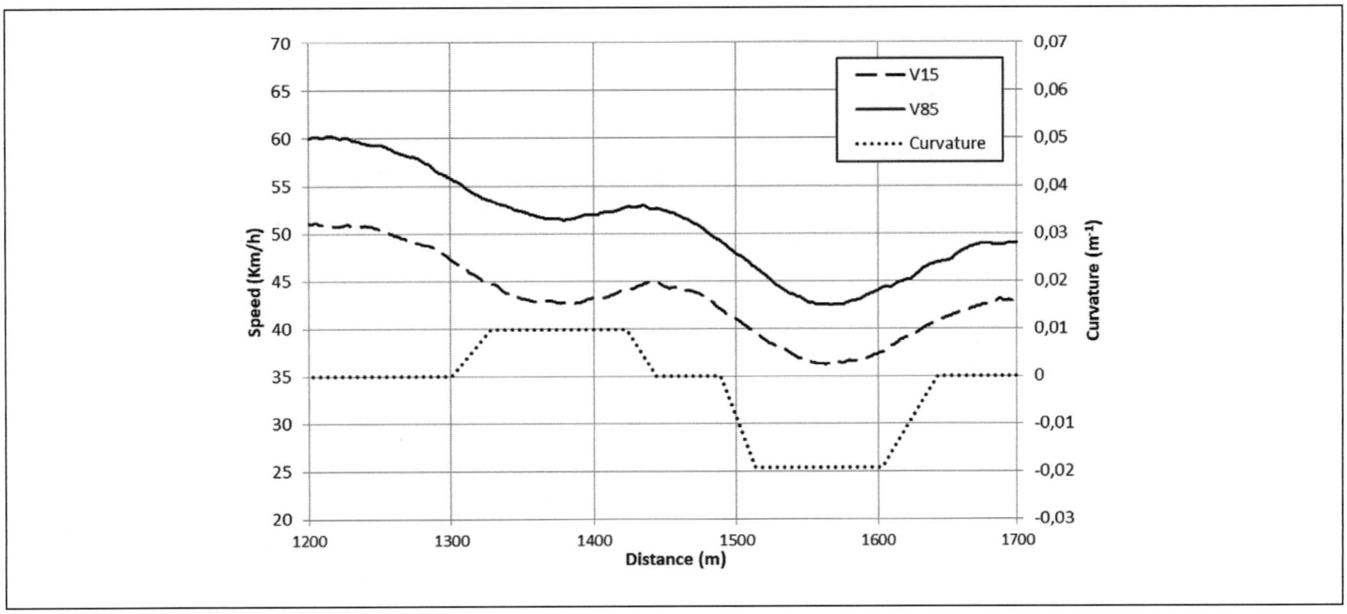

Figure 1. Horizontal curve selection.

obtained. From these profiles, the minimum speed and the speed at midcurve were identified for each horizontal curve.

The selection of horizontal curves for speed model calibration was based on the analysis of the speed profiles. Only horizontal curves that presented a significant speed reduction were selected, i.e., those that acted as a geometric control on drivers (Figure 1). This is only possible thanks to the use of continuous speed profiles. As a result, 105 horizontal curves were considered in this research.

Several geometric variables of the selected horizontal curves were studied. In the case of the horizontal alignment, the most relevant variables are the radius, the length and the deflection angle of the curve, the length of the preceding tangent, and the Curvature Change Rate (*CCR*), calculated as follows:

$$CCR = \frac{\gamma}{L} \quad (1)$$

where CCR = curvature change rate (gon/km); γ = deflection angle of the curve (gon); and L = curve length (km).

Regarding the vertical alignment, a preliminary analysis was performed to decide which variables should be included. For this, the grade at different points of the horizontal curve (at the point of curvature, at midpoint, and at the end) and the average grade were studied (Figure 2). The grade at the point of curvature corresponded to the grade along the preceding tangent. Only a few tangents presented different grades along them. In this case, the grade at the final part of the tangent was considered.

It was observed that the grade at the point of curvature and the grade at midpoint could better explain the phenomenon (Figure 2*a* and *b*). As expected, truck speeds decreased from a certain value of the grade. However, most of the selected horizontal curves presented a deceleration process that finished between these points, i.e., heavy vehicles adapted their speed before arriving at the midpoint (Figure 1). Therefore, heavy vehicle speeds were mainly influenced by the grade at the point of curvature, which was similar to the grade of the preceding tangent.

The trend associated with the grade at the end of the horizontal curve was counter-intuitive, i.e., the speeds became greater as grade increases (Figures 2*c*– unloaded trucks). Additionally, using the average grade along the horizontal curve might hide the true influence of the grade (Figure 2*d*).

So, the grade at the point of curvature was considered in this research.

Table 2 shows a statistical summary of the geometric variables of the selected horizontal curves.

The number of drivers required for each horizontal curve was also analyzed. For this, the following expression was used:

$$n = \frac{Z^2 \cdot \sigma^2}{e^2} \quad (2)$$

where n = number of drivers required; Z = quantile of a normal distribution considering a 95% confidence level (1.96); σ = speed deviation (km/h); and e = speed error (2 km/h). As a result, the number of drivers required in most of the selected horizontal curves was lower than 10

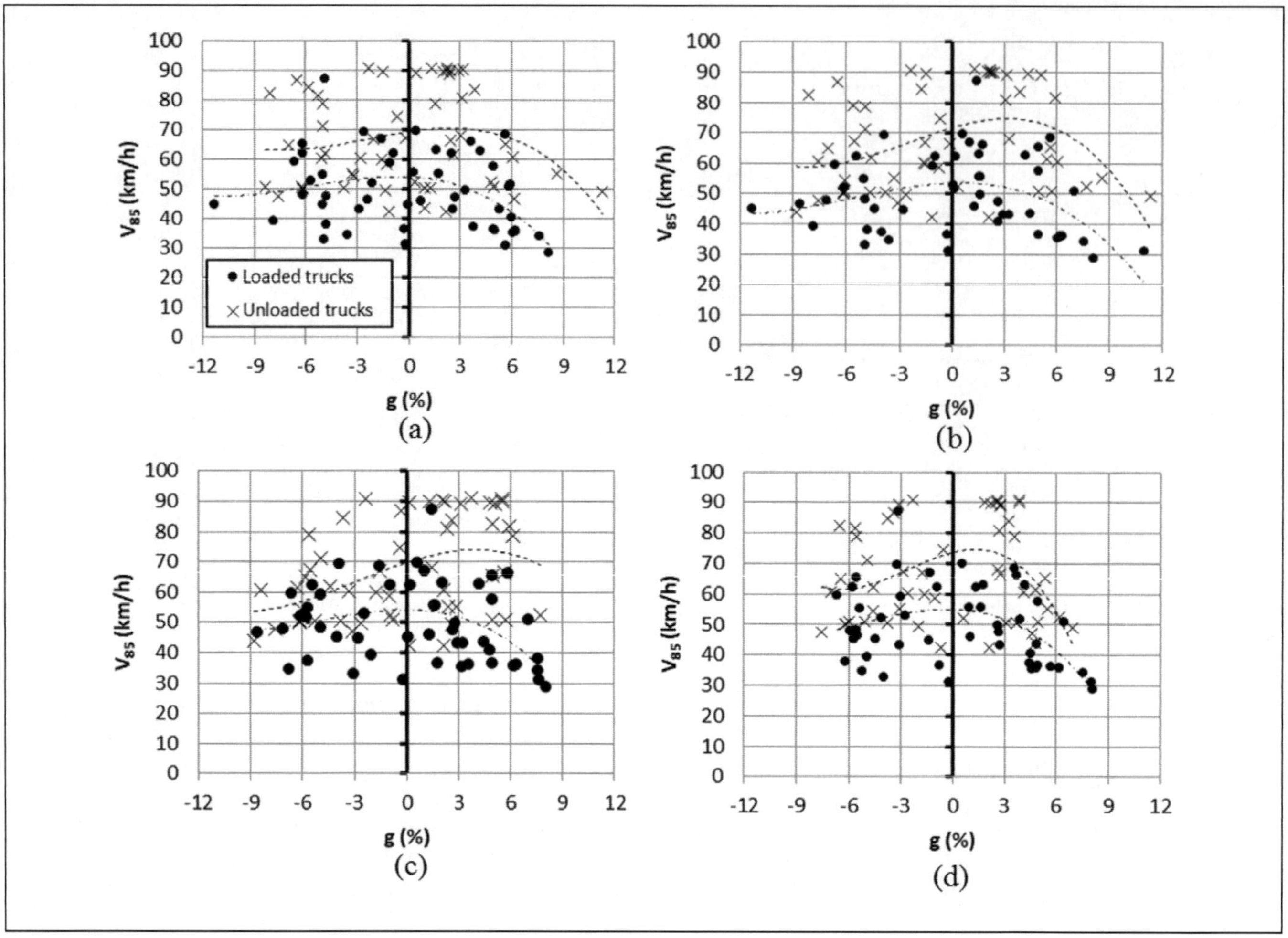

Figure 2. Analysis of the longitudinal grade.

Table 2. Statistical Summary of Geometric Variables at Horizontal Curves

Variable	Notation	Minimum	Maximum	Mean	Standard deviation
Radius (m)	R	18.45	1,178.36	167.43	228.52
Length of curve (m)	L	32.00	250.00	94.50	47.35
Deflection angle (gon)	γ	3.94	159.82	51.57	37.46
Length of preceding tangent (m)	L_t	7	1359	194	214
Curvature change rate (gon/km)	CCR	41.51	2,464.65	618.21	449.34
Grade (%)	g	−11.31	11.31	−0.04	4.68

mainly as a result of the low speed deviation observed on these locations. In this regard, the average speed deviation was 1.96 km/h. Even though, the average number of observations was around 80 drivers per horizontal curve.

Analysis

Most of the previous models for predicting the operating speed on horizontal curves were calibrated from spot speed data collected at midcurve, assuming the minimum speed was achieved at this location. However, these studies did not verify this hypothesis.

In this research, the minimum speed and the speed at midcurve were obtained from the continuous speed profiles collected. A hypothesis test based on the analysis of paired data was performed to determine whether the operating speed at midcurve (V_{85mc}) could be considered similar to the minimum operating speed (V_{85min}) or not.

For each horizontal curve, the following hypotheses were formulated: (a) Null hypothesis H_0: $V_{85mc}- V_{85min} = 0$; and (b) Alternative hypothesis H_1: $V_{85mc}- V_{85min} \neq 0$. The confidence level considered in the analysis was 95%.

As a result, the null hypothesis was rejected (t = 0,82536; p-*Value* = 0), i.e., V_{85mc} could not be considered similar to V_{85min} at a 95% confidence level. The same test was developed for the 15th percentile speed obtaining the same results.

Therefore, the hypothesis assumed by previous research should not be accepted without a preliminary analysis. In this way, the minimum speed was used for the calibration of speed prediction models.

85th Percentile Speed Model

Descriptive Analysis. The minimum operating speed (V_{85}) was identified for each horizontal curve from the 85th percentile speed profiles.

A preliminary correlation analysis was carried out to identify which horizontal geometric variables presented a greater influence on operating speed. This analysis found that the length of the preceding tangent and the length of the horizontal curve were weakly correlated with the operating speed. However, the radius of the horizontal curve, the Curvature Change Rate (*CCR*) and the deflection angle of the horizontal curve showed a great correlation with the operating speed. Specifically, the largest positive correlation coefficient was associated with the radius (0.5650), while the variables with the greatest negative correlation coefficient were associated with the *CCR* (0.7145) and the deflection angle (0.6058). Figure 3 shows these geometric variables versus the operating speed.

The trend of the data showed clearly two different populations, loaded and unloaded trucks. As expected, the operating speed for loaded trucks was lower than the operating speed for unloaded trucks. So, different regression analyses were performed.

On the other hand, the analysis of the relationship between the vertical alignment and the operating speed was focused on the grade. Although the grade was not a significant factor on downgrades, a decreasing trend was observed from a certain value of the grade (Figure 2).

To verify this phenomenon, the relationship between the horizontal and vertical alignment and the operating speed was studied jointly. Figure 3*e* shows the relationship between the radius, the grade, and the operating speed. In this case, the size of each point represents the grade value. It can be observed how the operating speed decreased as the grade increased for a given radius.

Modeling 85th Percentile Speed. Different regression models were calibrated based on the results described above. First, several speed models were developed considering different variables related to the horizontal alignment: radius, *CCR*, and deflection angle of the horizontal curve. Different functional forms were studied.

This analysis allowed the identification of the horizontal geometric variable that had a greater influence on the operating speed. The adjusted coefficient of determination was used as goodness of fit. Thus, the most accurate models for loaded and unloaded trucks were based on the radius of the horizontal curve as explanatory variable. These models are:

$$V_{85L} = 73.76 - 46.45/e^{0.0072 \cdot R} \quad R^2_{Adj} = 0.60 \tag{3}$$

$$V_{85U} = 87.55 - 60.37/e^{0.0102 \cdot R} \quad R^2_{Adj} = 0.74 \tag{4}$$

where V_{85L} = 85th percentile speed for loaded trucks (km/h); V_{85U} = 85th percentile speed for unloaded trucks (km/h); R = radius of the horizontal curve (m); and R^2_{Adj} = adjusted coefficient of determination.

The vertical alignment was introduced in the previous models to get more accurate speed models. For this, an analysis of the residuals as a function of the grade was carried out. This showed a homogeneous distribution of the residuals around 0 until a certain value of the grade, from which a decreasing linear trend was identified. Therefore, a compound model was proposed (Table 3).

It should be highlighted that the adjusted coefficients of determination increased significantly. This improvement was more obvious for loaded trucks, so the grade had more influence on these than on the unloaded. Thus, the vertical alignment had an important influence on the operating speed when the grade was greater than 4.23% for loaded trucks and 3.19% for unloaded trucks. On the contrary, the operating speed only depended on the horizontal alignment when the grade was lower than these values.

15th Percentile Speed Model

Descriptive Analysis. The calibration of the 15th percentile speed models was also based on the minimum speed along each horizontal curve.

The influence of the horizontal and vertical alignment on the 15th percentile speed was similar to that observed for the operating speed (Figure 3). The most influential variables related to the horizontal alignment were the radius, the *CCR*, and the deflection angle of the horizontal curve, which presented a correlation coefficient with respect to the 15th percentile speed equal to 0.5092, −0.6835, and −0.5734, respectively. Likewise, both loaded and unloaded truck speeds described a declining trend from a certain value of the grade.

Modelling 15th Percentile Speed. Several regression models were developed to assess the horizontal alignment

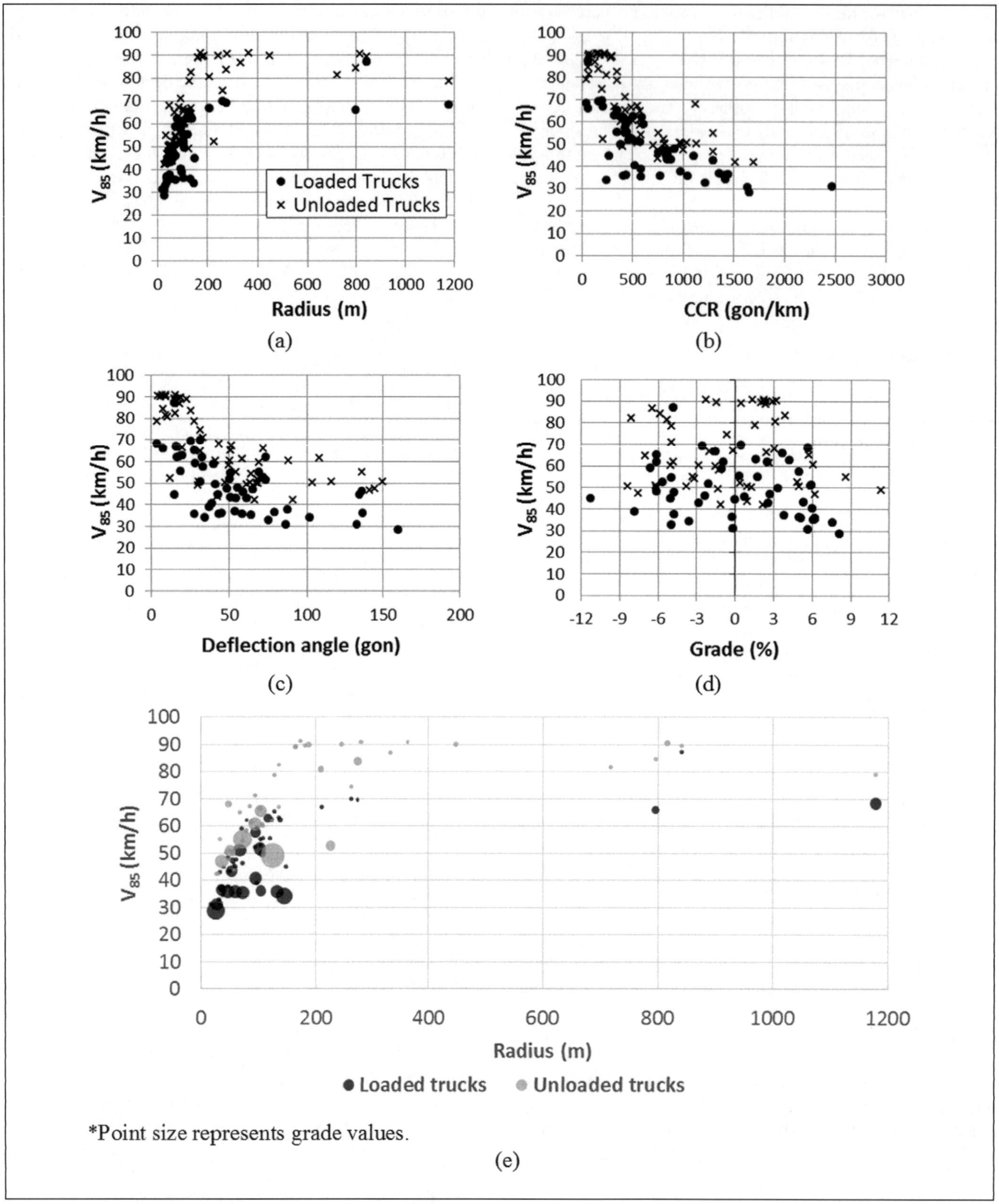

Figure 3. Geometric variables versus 85th percentile speed.

Table 3. 85th Percentile Speed Models

Operating speed models	
Loaded trucks	$V_{85L} = \begin{cases} 75.96 - \dfrac{44.56}{e^{0.00685 \cdot R}} & \text{if } g \leq 4.23\% \\ 75.96 - \dfrac{44.56}{e^{0.00685 \cdot R}} - 5.06 \cdot (g - 4.23) & \text{if } g > 4.23\% \end{cases} \quad R^2_{Adj} = 0.73$
Unloaded trucks	$V_{85U} = \begin{cases} 85.02 - \dfrac{60.62}{e^{0.01240 \cdot R}} & \text{if } g \leq 3.19\% \\ 85.02 - \dfrac{60.62}{e^{0.01240 \cdot R}} - 1.95 \cdot (g - 3.19) & \text{if } g > 3.19\% \end{cases} \quad R^2_{Adj} = 0.77$

where: V_{85L} = 85th percentile speed for loaded trucks (km/h); V_{85U} = 85th percentile speed for unloaded trucks (km/h); R = radius of the horizontal curve (m); g = grade (%); and R^2_{Adj} = adjusted coefficient of determination.

influence on 15th percentile speed. Different functional forms were studied and the adjusted coefficient of determination was given in all regressions as goodness of fit.

As a result, Equations 5 and 6 show the most accurate models, which used the radius of the horizontal curve as explanatory variable.

$$V_{15L} = 61.00 - 38.93/e^{0.0080 \cdot R} \quad R^2_{Adj} = 0.54 \tag{5}$$

$$V_{15U} = 76.66 - 56.72/e^{0.0108 \cdot R} \quad R^2_{Adj} = 0.69 \tag{6}$$

where V_{15L} = 15th percentile of the distribution of speeds for loaded trucks (km/h); V_{15U} = 15th percentile of the distribution of speeds for unloaded trucks (km/h); R = radius of the horizontal curve (m); and R^2_{Adj} = adjusted coefficient of determination.

In addition, some speed models were calibrated considering both horizontal and vertical alignment. For this, the residuals of the previous models were analyzed as a function of the grade. In this regard, a decreasing linear trend was identified from a certain value of the grade. Therefore, the same functional form used for the operating speed was proposed (Table 4).

As a result, the grade showed an important influence on speed from 3.14% and 3.06% for loaded and unloaded trucks, respectively. As expected, introducing the vertical alignment in the models improved their accuracy, since the adjusted coefficient of determination increased significantly.

Discussion

Most previous studies about heavy vehicle speeds presented deficient data collection because of the number of locations, the number of observations, and the data collection methodology. In this regard, this research has been developed from continuous speed profiles collected along a large amount of horizontal curves.

These continuous speed profiles allowed the analysis of the hypothesis that there is not a significant difference between the speed at midcurve and the minimum speed along the horizontal curve. As a result, this hypothesis was rejected and the minimum speed was found to be statistically lower than the speed at midcurve at a 95% confidence level.

The most influential variable on truck speeds was the radius. This result was consistent with the outcomes obtained in previous research (*3, 7, 18*). Regarding this, the influence of the radius on heavy vehicle speeds was different depending on whether the vehicle was loaded or not. For unloaded trucks, the radius was dominant up to values around 200 m. This means that for a radius lower than 200 m, a small variation of the value of the radius leads to a great speed variation. Conversely, a great variation of the value of the radius was needed to observe a significant speed variation when the radius was greater than 200 m. However, the influence of the radius was greater for loaded trucks, since a high influence was identified up to a radius of 300 m. This is due to loaded trucks having a higher center of gravity.

Despite identifying the same trend according to the radius of the horizontal curve, the new operating speed models estimated greater speeds than the model developed by Jacob and Anjaneyulu (*7*). This phenomenon was probably related to the difference between the Spanish and Indian two-lane rural roads, which present different posted speed limits and types of heavy vehicles. In this context, the trucks considered in this research might have a lower weight-to-power ratio (WPR) than the heavy vehicles considered by Jacob and Anjaneyulu (Figure 4).

It should be noted that the calibrated model estimates higher speeds for loaded trucks than for unloaded trucks for very low radii ($R < 20$ m). Thus, these models should be used for radii greater than 20 m.

On the other hand, the grade at the point of curvature also showed a significant influence on truck speeds.

Table 4. 15th Percentile Speed Models

Loaded trucks	$V_{15L} = \begin{cases} 64.17 - \frac{37.24}{e^{0.00720 \cdot R}} & \text{if } g \leq 3.14\% \\ 64.17 - \frac{37.24}{e^{0.00720 \cdot R}} - 3.28 \cdot (g - 3.14) & \text{if } g > 3.14\% \end{cases}$ $R^2_{Adj} = 0.69$
Unloaded trucks	$V_{15U} = \begin{cases} 76.74 - \frac{57.58}{e^{0.01185 \cdot R}} & \text{if } g \leq 3.06\% \\ 76.74 - \frac{57.58}{e^{0.01185 \cdot R}} - 2.43 \cdot (g - 3.06) & \text{if } g > 3.06\% \end{cases}$ $R^2_{Adj} = 0.74$

where V_{15L} = 15th percentile speed for loaded trucks (km/h); V_{15U} = 15th percentile speed for unloaded trucks (km/h); R = radius of the horizontal curve (m); and g = grade (%); and R^2_{Adj} = adjusted coefficient of determination.

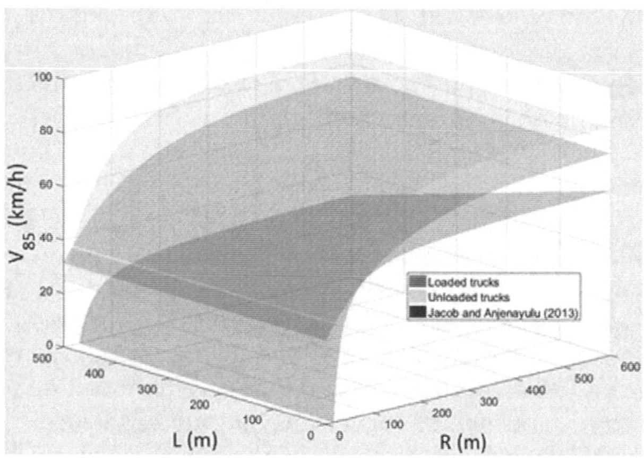

Figure 4. 85th percentile speed models versus Jacob and Anjaneyulu's model.

Contrary to expectations, the grade threshold was lower for unloaded trucks (3%) than for loaded ones (4%) according to the 85th percentile speed models. Regarding the 15th percentile speed models, heavy vehicle speeds decreased for grades greater than 3% for both loaded and unloaded trucks. However, the influence of the grade was higher for loaded trucks than for unloaded trucks, since the regression coefficients related to loaded trucks were greater.

Other authors pointed out that the heavy vehicle speeds decreased as the grade increased, but not from a certain grade value (*3*, *18*). These conclusions might be related to the geometric characteristics of the horizontal curves under study. Specifically, Morris and Donnell (*18*) studied truck speeds on multilane highways, which usually present greater radius and tangent lengths. In this situation, drivers are not as influenced by the horizontal alignment as on two-lane rural roads, where this alignment predominates over the vertical alignment on downgrades.

According to Donnell et al. (*3*), the operating speed at midcurve was also influenced by the length and grade of the departure tangent, which led this study to conclude that the grade at the point of curvature significantly influenced truck speeds, which was similar to the grade of the preceding tangent for most of the horizontal curves. So, the geometric and operating characteristics of the approach tangent have an important effect on horizontal curve speed, which requires further research.

Finally, a comparison between 85th and 15th percentile speed models was carried out (Figure 5). The difference between both speed percentiles became lower as the radius decreased for both loaded and unloaded trucks, i.e., sharper geometries produced lower speed deviations. In addition, the difference between both speed percentiles was lower for unloaded trucks than for loaded ones. As an example, this speed difference was approximately 8 km/h for unloaded trucks considering the high values of the radius, whereas this speed difference increased to 12 km/h for loaded trucks. Therefore, loaded trucks underwent a greater speed deviation than unloaded ones discounting the effects of the vertical alignment.

Regarding vertical alignment, a different trend was observed depending on whether the vehicle was loaded or not. For loaded trucks, the difference between both speed percentiles was lower as the grade increased. On the contrary, the speed difference increased as the grade increased for unloaded trucks. In this way, loaded trucks presented a lower speed deviation than unloaded trucks just focusing on the vertical alignment.

Thus, loaded trucks were more influenced by the vertical alignment and unloaded trucks were more influenced by the horizontal alignment.

Conclusions and Further Research

The development of truck speed profiles is crucial to assess the interaction between passenger cars and heavy vehicles, which is a critical factor in road safety.

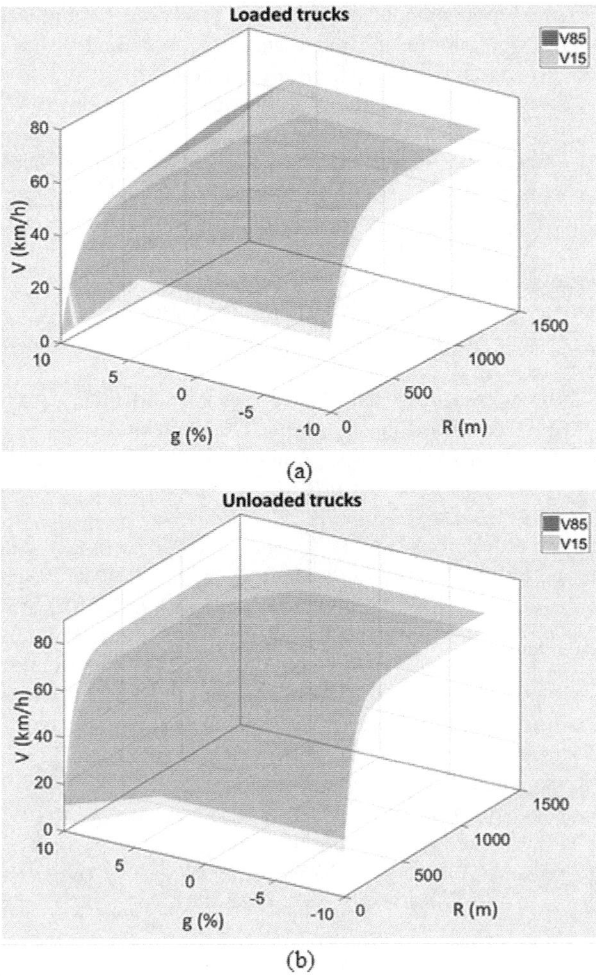

Figure 5. Comparison between 85th and 15th percentile speed models.

However, this phenomenon is rarely considered due to the lack of truck speed models.

This study presents several speed models, which include geometric characteristics as explanatory variables to predict truck speeds on horizontal curves of two-lane rural roads. These models were calibrated based on the minimum speed identified from continuous speed profiles collected using GPS tracking devices.

The results showed the combined effect of the horizontal and vertical alignment on truck speeds. The most influential variables were the radius of the horizontal curve and the grade at the point of curvature. In addition, two different trends were identified which were related to loaded and unloaded trucks.

The influence of the vertical alignment was only observed on upgrades. In this way, operating speeds decreased for grades greater than 4% and 3% for loaded and unloaded trucks, respectively. In addition, the influence of the grade was greater for loaded trucks than for unloaded trucks.

85th and 15th percentile speed models were calibrated for both loaded and unloaded trucks. These models were compared with those developed previously. Although the same trends were observed, the use of the proposed models are only recommended on Spanish two-lane rural roads due to the different characteristics of heavy vehicles among countries.

The models presented in this research are part of the development of a new operating speed profile for heavy vehicles. The continuous speed profiles collected in this research will be used to develop speed models on tangents and to analyze tangent-to-curve speed variations, and acceleration and deceleration rates. This will allow the completion of the truck operating speed profile model. In this way, the interaction between passenger cars and heavy vehicles will be able to be considered in the analysis of the geometric design consistency.

Acknowledgments

The study presented in this paper is part of the research project titled "CASEFU - Estudio experimental de la funcionalidad y seguridad de las carreteras convencionales" (TRA2013-42578-P), subsidized by the Spanish Ministry of Economy, Industry, and Competitiveness and the European Social Fund. In addition, the authors would like to thank the companies SAV, SAEVI, and OAM, for their cooperation in field data gathering.

Author Contributions

The authors confirm contributions to the paper as follows: study conception and design: Llopis-Castelló, D., Pérez-Zuriaga, A. M., and García, A.; data collection: Llopis-Castelló, D., Pérez-Zuriaga, A. M., and García, A; analysis and interpretation of results: Llopis-Castelló, D., Pérez-Zuriaga, A. M., and González-Hernández, B; draft manuscript preparation: Llopis-Castelló, D. and González-Hernández, B. All authors reviewed the results and approved the final version of the manuscript.

References

1. World Health Organization. *Global Status Report on Road Safety 2015*. World Health Organization, Switzerland, 2015.
2. American Association of State Highway and Transportation Officials. *A policy on geometric design of highways and streets*, 6th ed. American Association of State Highway and Transportation Officials, Washington, DC, 2011.
3. Donnell, E., Y. Ni, M. Adolini, and L. Elefteriadou. Speed Prediction Models for Trucks on Two-lane Rural Highways. *Transportation Research Record: Journal of the Transportation Research Board*, 2001. 1751: 44–55.
4. Harwood, D. W., D. J. Torbic, K. R. Richard, W. D. Glauz, and L. Elefteriadou. *Review of Truck*

Characteristics as Factors in Roadway Design. Report 505. Transportation Research Board of the National Academies, Washington, D.C., 2003.

5. Leisch, J. E., and J. P. Leisch. New Concepts in Design-speed Application. *Transportation Research Record*: Journal of the Transportation Research Board, 1977. 631: 4–14.
6. Misaghi, P., and Y. Hassan. Modeling Operating Speed and Speed Differential on Two-lane Rural Roads. *Journal of Transportation Engineering*, Vol. 131, No. 6, 2005, pp. 408–418.
7. Jacob, A., and M. V. L. R. Anjaneyulu. Operating Speed of Different Classes of Vehicles at Horizontal Curves on Two-lane Rural Highways. *Journal of Transportation Engineering*, Vol. 139, No. 3, 2012, pp. 287–294.
8. Gibreel, G. M., S. M. Easa, and I. A. El-Dimeery. Prediction of Operating Speed on Three-Dimensional Highway Alignments. *Journal of Transportation Engineering*, Vol. 127, No. 1, pp. 21–30.
9. Saifizul, A. A., H. Yamanaka, and M. R. Karim. Empirical Analysis of Gross Vehicle Weight and Free Flow Speed and Consideration on its Relation with Differential Speed Limit. *Accident Analysis & Prevention*, Vol. 43, No. 3, 2011, pp. 1068–1073.
10. Bester, C. Truck Speed Profiles. *Transportation Research Record: Journal of the Transportation Research Board*, 2000. 1701: 111–115.
11. Lan, C. J., and M. Menendez. Truck Speed Profile Models for Critical Length of Grade. *Journal of Transportation Engineering*, Vol. 129, No. 4, 2003, pp. 408–419.
12. Rakha, H., and B. Yu. Truck Performance Curves Reflective of Truck and Pavement Characteristics. *Journal of Transportation Engineering*, Vol. 130, No. 6, 2004, pp. 753–767.
13. Echaveguren, T., and D. Carrasco. Estimación de longitud crítica en pendientes ascendentes en caminos bidireccionales [Estimation of critical length on ascending slopes on bidirectional roads]. *Revista ingeniería de construcción* [*Engineering Magazine of Construction*], Vol. 30, No. 1, 2015, pp. 5–16.
14. Camacho-Torregrosa, F. J., A. M. Pérez-Zuriaga, J. M. Campoy-Ungría, A. García, and A. P. Tarko, Use of Heading Direction for Recreating the Horizontal Alignment of an Existing Road. Computer-Aided Civil and Infrastructure Engineering, Vol. 30, No. 4, 2015, pp. 282–299.
15. Plan Nacional de Ortofotografía Aérea. Ministerio de Fomento, España, 2016. http://pnoa.ign.es/caracteristicas-tecnicas
16. Zuriaga, A., A. García, F. Torregrosa, and P. D'Attoma. Modeling Operating Speed and Deceleration on Two-lane Rural Roads with Global Positioning System Data. *Transportation Research Record: Journal of the Transportation Research Board*, 2010. 2171: 11–20.
17. Pérez-Zuriaga, A. M., F. J. Camacho-Torregrosa, A. García, and J. M. Campoy-Ungría. Application of Global Positioning System and Questionnaires Data for the Study of Driver Behaviour on Two-lane Rural Roads. *IET Intelligent Transport Systems*, Vol. 7, No. 2, 2013, pp. 182–189.
18. Morris, C. M., and E. T. Donnell. Passenger Car and Truck Operating Speed Models on Multilane Highways with Combinations of Horizontal Curves and Steep Grades. *Journal of Transportation Engineering*, Vol. 140, No. 11, p. 04014058, 2014.

The Standing Committee on Operational Effects of Geometrics (AHB65) peer-reviewed this paper (18-00823).

Article

TRR
JOURNAL OF THE TRANSPORTATION RESEARCH BOARD

Transportation Research Record
2018, Vol. 2672(17) 83–95

DOI: 10.1177/0361198118784384
journals.sagepub.com/home/trr
SAGE

Studying Driving Behavior on Horizontal Curves using Naturalistic Driving Study Data

Bashar Dhahir[1] and Yasser Hassan[1]

Abstract

Many studies have been conducted to develop models to predict speed and driver comfort thresholds on horizontal curves, and to evaluate design consistency. The approaches used to develop these models differ from one another in data collection, data processing, assumptions, and analysis. However, some issues might be associated with the data collection that can affect the reliability of collected data and developed models. In addition, analysis of speed behavior on the assumption that vehicles traverse horizontal curves at a constant speed is far from actual driving behavior. Using the Naturalistic Driving Study (NDS) database can help overcome problems associated with data collection. This paper aimed at using NDS data to investigate driving behavior on horizontal curves in terms of speed, longitudinal acceleration, and comfort threshold. The NDS data were valuable in providing clear insight on drivers' behavior during daytime and favorable weather conditions. A methodology was developed to evaluate driver behavior and was coded in Matlab. Sensitivity analysis was performed to recommend values for the parameters that can affect the output. Analysis of the drivers' speed behavior and comfort threshold highlighted several issues that describe how drivers traverse horizontal curves that need to be considered in horizontal curve design and consistency evaluation.

Many models are available in the literature to predict operating speed and other driver behavior parameters on horizontal curves based on the highway's geometric features (*1*). These models serve different purposes including design consistency evaluation or horizontal curve design. However, the reliability of such models can be compromised if the collected data do not reflect the actual drivers' behavior due to limitations in the data collection equipment and methodologies. As it has been stated, "researchers too often have resorted to gathering the data that are available instead of those needed to address the salient questions" (*2*). For example, a methodology targeting speed or lateral acceleration measurements at specific locations, such as beginning or middle of horizontal curves, may not capture the minimum or maximum values along the curve. Also, data processing and aggregation can cause loss of information related to the distribution of data (*3*). The data collection might also be associated with inherent issues that compromise the accuracy of collected data, and hence the developed models. Examples are potential human errors in manual speed data collection; cosine error in radar/laser speed data collection; unrepresentative or small samples in simulator experiments; and driver's heightened alertness in short-term instrumented vehicle experiments (*4*). Using data extracted from naturalistic driving studies (NDS) is a promising alternative to collect realistic driver behavior data and overcome these issues. It could provide data for many variables for reasonable duration, different roadway characteristics, and different driving and environmental conditions. Combining the recorded driving behavior data with roadway geometry, traffic, and weather conditions would be a powerful tool to describe drivers' behavior on horizontal curves. Processing the substantial amounts of NDS data by third-party researchers (not involved in the data collection) would benefit significantly from a detailed procedure for extracting the relevant measures describing driver behavior.

[1]Department of Civil and Environmental Engineering, Carleton University, Ottawa, Ontario, Canada

Corresponding Author:
Address correspondence to Yasser Hassan: yasser.hassan@carleton.ca

Objectives

The main objective of this paper is to study driving behavior on horizontal curves in terms of speed, longitudinal acceleration, and lateral acceleration using NDS data. The ability to quantitatively evaluate these driver behavior parameters would help identify the critical locations and design parameters on horizontal curves and develop more reliable models to predict maximum and minimum speed, speed reduction, longitudinal acceleration, and lateral acceleration. Thus, reliable models can be developed for operating speed profiles, reliability analysis of horizontal curves (*5*), and design consistency evaluation (*6*). To achieve this objective, a methodology is developed to process data files of NDS driver trips to evaluate the different measures quantifying driver behavior on horizontal curves. Frequency of these parameters are determined at different areas of the approach tangent and curve and are examined to determine the driving patterns on horizontal curves.

Methodology

This paper focuses on using the NDS data, conducted through the second Strategic Highway Research Program (SHRP 2), to evaluate driver behavior related to speed, longitudinal acceleration, and lateral acceleration on horizontal curves. This NDS involved recording up to three years of all trips for more than 3,100 volunteer drivers in six states in the USA, which makes it the largest NDS to date (*7*). The data acquisition system installed in the vehicles recorded continuous data of vehicle speed, acceleration, braking, vehicle controls, lane position, forward radar, and video views to the front and rear of the vehicle and on the driver's face and hands.

The methodology followed in this paper comprises the following tasks: site selection and data collection, removing unusable data, preliminary data processing, developing a procedure to determine the driver behavior measures, and sensitivity analysis to evaluate the effect of any assumed parameters on the driver behavior measures.

Site Selection and Data Collection

Horizontal curve sites selected for this study were all located in Washington State (one of the states participating in the NDS). Presence of different terrain types, varying weather conditions throughout the year, and availability of road geometric data through the Washington State Department of Transportation (WSDOT) and Highway Safety Information System (HSIS) were the main reasons behind this selection. Data were screened to identify appropriate horizontal curves on two-lane rural arterial highways, resulting in the selection of 21 and 25 curves in rolling and mountainous terrain, respectively. Each selected curve is preceded by an independent tangent, which should be long enough for vehicles to reach their desired speeds independently from the curvature on either side of the tangent. Several researchers, such as Fitzpatrick et al. (*8*), reported that 200 m is the minimum tangent length to act as an independent tangent. A total of 9 and 21 curves in rolling and mountainous terrain, respectively, were preceded by an independent tangent in one direction, which was the only travel direction considered in this study. Each of the remaining 12 and 4 curves in rolling and mountainous terrain were preceded by independent tangents in both directions, and therefore each travel direction was considered as a separate curve. Thus, the NDS dataset for this study consisted of 33 and 29 curves in rolling and mountainous terrain, respectively. The geometric parameters of these curves are shown in Table 1.

Preliminary Data Processing

The total number of trips extracted from the NDS was 3,503 and 2,731 for rolling and mountainous terrains, respectively. These trips were driven by 178 participants of different ages and professions using four types of light vehicles: passenger car, sports utility vehicle, van/minivan, and pickup/light truck vehicles. Each trip in the NDS dataset has its own file, which includes variables such as location of the vehicle (GPS latitude and longitude), longitudinal and lateral acceleration, speed, and headway (in meters and seconds). However, 15% to 18% of the trips were removed from the dataset because they lacked essential data such as location or speed, reducing the number of trips to 2,874 and 2,316 for rolling and mountainous terrains, respectively.

The trips were further screened based on trip time and weather conditions. A trip was classified as daytime if it occurred after the civil twilight in the morning and before the civil twilight in the evening, where civil twilight is the instant in the morning or evening when the center of the sun is at a depression angle of 6° below an ideal horizon (*9*). Daily weather conditions data for three years (2011–2013) of seven selected weather stations were obtained from the Office of Washington State Climatologist, where each curve was approximately 2 to 9 km away from at least one station. The total number of trips during daytime and favorable weather conditions in rolling and mountainous terrain were 1,952 and 1,816, respectively. The lowest number of unique drivers and total trips on any curve was 30 and 34, respectively. Because only a few trips were available in other conditions, only daytime and favorable weather conditions trips were considered in this paper.

Table 1. Geometric Data of Study Curves

Curve ID	R (m)	L (m)	e	TD	Curve ID	R (m)	L (m)	e	TD
RT 01	218	257	0.080	L	MT 01	348	511	0.06	L
RT 02	437	186	0.070	L	MT 02	289	319	0.076	L
RT 03	218	158	0.080	L	MT 03	349	252	0.06	R
RT 04	218	158	0.080	R	MT 04	582	496	0.032	R
RT 05	291	421	0.068	L	MT 05	635	763	0.045	L
RT 06	388	276	0.068	R	MT 06	291	320	0.072	L
RT 07	388	276	0.068	L	MT 07	582	424	0.053	L
RT 08	291	297	0.068	L	MT 08	698	507	0.06	R
RT 09	291	297	0.068	R	MT 09	698	507	0.06	L
RT 10	318	261	0.060	R	MT 10	279	201	0.096	L
RT 11	318	261	0.060	L	MT 11	303	381	0.07	R
RT 12	274	254	0.080	R	MT 12	283	308	0.06	R
RT 13	274	254	0.080	L	MT 13	349	277	0.067	R
RT 14	274	164	0.080	L	MT 14	349	277	0.067	L
RT 15	274	164	0.080	R	MT 15	332	124	0.06	L
RT 16	146	120	0.080	L	MT 16	279	116	0.096	L
RT 17	174	299	0.100	L	MT 17	410	587	0.068	L
RT 18	297	313	0.080	R	MT 18	436	270	0.062	L
RT 19	250	326	0.073	R	MT 19	332	225	0.06	L
RT 20	349	271	0.060	L	MT 20	436	158	0.061	L
RT 21	349	262	0.092	L	MT 21	436	158	0.06	R
RT 22	349	262	0.092	R	MT 22	698	442	0.055	L
RT 23	291	174	0.068	L	MT 23	698	442	0.037	R
RT 24	291	174	0.068	R	MT 24	434	258	0.073	L
RT 25	318	298	0.050	R	MT 25	582	244	0.058	L
RT 26	318	298	0.050	L	MT 26	696	371	0.053	R
RT 27	250	177	0.073	R	MT 27	388	488	0.086	R
RT 28	250	177	0.073	L	MT 28	498	303	0.06	L
RT 29	349	445	0.090	L	MT 29	873	161	0.039	R
RT 30	349	445	0.090	R					
RT 31	218	196	0.080	R					
RT 32	291	165	0.068	L					
RT 33	291	165	0.068	R					

Note: R = curve radius; L = curve length; e = superelevation rate; TD = turning direction; RT = rolling terrain; MT = mountainous terrain; R = right turn; L = left turn.

The final trip data were matched with the roadway information database to determine the points in the trip file that correspond to beginning and end of curve (BC and EC). The GPS data and BC/EC coordinates were used to determine the points where the vehicle is closest to BC/EC. Because speed data were available every 0.1 s while the GPS data were available every 1.0 s, the vehicle locations at BC/EC were refined by interpolating the location based on vehicle speed and time stamp. Another challenge was to interpret the vehicle headway data. According to Virginia Tech Transportation Institute (VTTI), the trip data would not include headway information if no objects were detected ahead of the host vehicle at the maximum detection range of 200 m or if detected objects were not classified as a lead vehicle. Therefore, points where data were available except for headway information were classified as free-flow conditions, if sensors were sampling at same rate.

Procedure to Evaluate Driver Behavior

General Speed State. To evaluate a driver's speed and longitudinal acceleration behavior on curve and approach tangent, it is important to consider the acceleration and speed profiles. As shown in Figure 1, because of the high frequency of speed and acceleration measurements (10 readings/s), an average acceleration based on the instantaneous measurements can be misleading. For example, by dividing the curve in Figure 1 into three equal-length zones, taking the average of the acceleration values at the middle zone would indicate a "deceleration" state; whereas the speed profile shows a general "acceleration" trend. In addition, the high frequency of speed measurements can experience an instantaneous reduction in speed (deceleration) while the general trend indicates acceleration, or vice versa. A likely reason for this discrepancy is the effect of road grade on the accelerometer

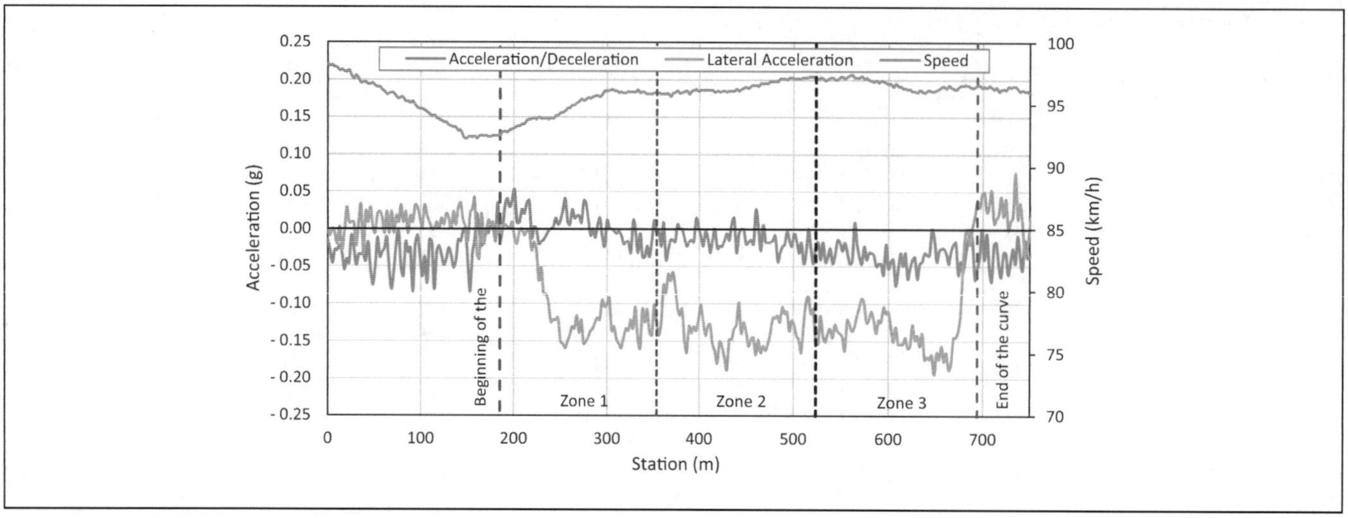

Figure 1. Example of profiles of speed and longitudinal and lateral accelerations.

used to measure longitudinal acceleration. Therefore, the general speed trend or state of a specific trip should be evaluated based on the general trend of speed profile using multiple speed measurements.

Furthermore, the speed trend can vary depending on the vehicle's relative location on the curve. Therefore, an analysis of general speed should be performed on curve segments or zones that can capture the general speed trend at various locations of the curve. Subsequently, a procedure was developed and coded in Matlab to divide the curve into a user-specified number of zones and determine the general speed state in each zone. The procedure involves the following steps:

1. Because of the different frequencies of speed and GPS measurements, the GPS data do not indicate the locations of all speed measurements. Therefore, the location of each speed measurement was determined using cumulative numerical integration of speed over time. Because of the small time-interval between each two speed measurements (0.1 s), cumulative trapezoid integration was deemed satisfactory. Additional distance was added for the first or last speed measurement if either point did not coincide with BC or EC, respectively, due to a gap in speed data. This distance was calculated as the speed at the first (or last) point multiplied by the time-gap to BC (or EC). Accuracy of this calculation increases as the speed-data-gap decreases.
2. The curve is divided into zones of equal length and the speed measurements are assigned to the proper zone based on their calculated locations on the curve. It was found that some trips lacked speed measurements for a considerable length of one or more zones. If a trip was missing speed measurements on the first or last half of any zone length, it was thought that the trend of the remaining measurements might not accurately reflect the general speed trend, and the trip was deemed unusable for the analysis of speed state.
3. Linear regression was used to fit the speed measurements at each zone to a straight line versus time. Thus, the slope of speed line would represent the average trip acceleration in each zone. Subsequently, the trip speed state in each zone is classified as "acceleration", "deceleration," or "constant" if the speed line has a positive, negative, or zero slope, respectively. For practical reasons, the speed state was considered "constant" when the absolute value of the slope did not exceed a specific acceleration threshold (accThreshold) or if the slope as a regression coefficient was statistically insignificant (i.e. statistically not different from zero) at a specific level of significance threshold (pvalThreshold). The values assigned to these two parameters can affect the results of speed state.

Curve Speed Measures. The developed Matlab code also determines several measures of trip speed including speed on approach tangent; speed at beginning, middle, and end of curve; and minimum and maximum speeds on the curve ($V_{crv-min}$ and $V_{crv-max}$) and their locations. In this paper, the analysis focuses on the locations of $V_{crv-min}$ and $V_{crv-max}$ to examine the assumption of constant curve speed in most speed profile models, including the Interactive Highway Safety Design Model (IHSDM) (*8*). For easy display and aggregation of the results, the code divides the curve into segments or bins of equal length and determines the frequencies of occurrence of $V_{crv-min}$

and $V_{crv-max}$ in each bin. Because a trip can have multiple occurrences of $V_{crv-min}$ and/or $V_{crv-max}$, the frequencies presented in this paper correspond to all occurrences of $V_{crv-min}$ and $V_{crv-max}$ of each trip. The results of this part of the procedure and Matlab code do not depend on any user-specified parameter, other than the number of bins, and can be determined for any trip with any number of speed measurements. However, it was decided to exclude the trips that were missing speed measurements deeming them unusable for the general speed state.

Driver Comfort Threshold. Driver comfort is a human factor issue related to "kinesthesia", which is defined as "the awareness of the orientation and the rates of moment of different parts of the body arising from stimulation of receptors in the joints, muscles, and tendons" (*10*). When a vehicle traverses a curved path, lateral acceleration is developed, whose magnitude depends on the vehicle speed and curve radius. The combined effects of lateral acceleration and the vehicle's roll angle is a driver's sensation of forces pushing them to the outside of the curve. If the lateral acceleration exceeds a specific limit, the driver would reduce the speed on the curve to mitigate the effect of the developed lateral acceleration. This limit is commonly known as the "driver comfort threshold" (*5*), which varies from one driver to another. The current design guides adopt these comfort thresholds to set the design values of lateral friction coefficient in designing horizontal curves.

To determine comfort thresholds, it is important to consider the dynamic phases of a vehicle when negotiating a horizontal curve. As seen in Figure 1, the transient turn-entry phase starts at the beginning of the curve, where the lateral velocity (component of velocity in the direction of the lateral axis) develops from zero and changes with time. The second phase is the steady state cornering phase, where the vehicle moves on the curve with almost constant radius and lateral velocity. Finally, the transient turn-exit phase occurs at the end of the curve where the lateral velocity changes with time to return to zero (*11*). The lateral acceleration has different values in these phases, with extreme values experienced usually in the transient phases. However, because of the very short driver exposure time to these extreme values, they will not likely prompt driver's speed adjustment and would not reflect the comfort threshold. Alternatively, the comfort threshold should correspond to the almost-constant value of lateral acceleration at the steady state cornering phase. Said et al. (*12*) calculated comfort threshold for each driver on a specific curve as the average lateral acceleration after removing extreme observations. However, an average of all readings on the curve might not catch the actual comfort threshold for each driver unless the driver maintains a constant speed along the entire curve.

An alternative methodology was developed in this paper and coded in Matlab to evaluate the comfort threshold for each trip in the dataset by searching for the minimum value of average lateral acceleration measurements while the trip is at a constant speed state. To find this minimum, the user specifies the minimum length and minimum number of speed points to be used to establish this speed state. The larger distance based on these two parameters is set as a sliding window along the curve. Linear regression is used to find the speed state for each window location using the procedure explained earlier. If the speed state is "constant," the average lateral acceleration at that window is calculated and compared to any previous value for this trip and the minimum value is kept. At the end of the loop, the comfort threshold is determined as the minimum value of average lateral acceleration while the trip is at a state of constant speed. The code also determines the minimum and mean speeds corresponding to the window corresponding to the comfort threshold.

The output of this procedure depends on the parameters accThreshold and pvalThreshold, which are used to establish the speed state. In addition, the output depends on the size of sliding window. A large window is expected to increase the value of comfort threshold because of including speed points beyond where the minimum speed is reached. On the other hand, a very small window can deem many trips to be unusable if a gap equal to the window size is found in the speed data or if short-term speed fluctuations cause the small window to always be in "acceleration" or "deceleration" speed state.

Sensitivity Analysis

A sensitivity analysis was performed using the rolling terrain curves to assess the effect of three main parameters specified in the Matlab code on the results, which are pvalThreshold, accThreshold, and minimum size of sliding window (minSSWin). All these parameters would affect the calculations of comfort threshold while the general speed state is affected by pvalThreshold and accThreshold only. It should be noted that selecting the number of zones to examine driver acceleration behavior on a specific curve should depend on the curve length. However, to compare and aggregate the results on different curves, the same number of zones is used for all curves in this paper. As shown in Table 1, around 85% of the rolling terrain curves had a length less than 300 m. Therefore, the general speed states were evaluated by dividing each curve to three zones. Thus, each zone length was less than 100 m for most curves.

The first sensitivity analysis was run to examine the effect of pvalThreshold using accThreshold of 0.1 m/s^2, minSSWin of 0.3, and three values of pvalThreshold of

0.01, 0.05, and 0.1. As shown in Figure 2*a*, the percentages of trips for the three speed states corresponding to all curves are almost constant for all values of pvalThreshold. In most cases, the maximum change in trip percentage for a specific state in a specific zone was between 2% and 3%. The largest changes were less than 5.5% and corresponded to the constant state which corresponded to over 50% of the trips. The same trend was observed for each curve individually. As for the effect on comfort threshold, the absolute change in mean comfort threshold on each curve was less than 0.01 g, and the percentage change for most curves was less than 2% (Figure 2*c*). Similar findings were observed in relation to the standard deviation of comfort threshold where the maximum absolute change was approximately 0.005 g, and the percentage change for most curves was less than 5%. Therefore, the results are not sensitive to pvalThreshold, and a 5% value was selected for the rest of this paper.

The sensitivity of accThreshold was analyzed using five values ranging from 0.1 to 0.5 m/s^2, while setting pvalThreshold as 0.05 and minSSWin as 0.3. The results of mean and standard deviation of comfort threshold showed lack of sensitivity to accThreshold. The maximum difference in mean and standard deviation values on any curve was 0.023 g and 0.018 g, respectively. As for the general speed state, Figure 2*b* shows that the percentages of trips depend on the value of accThreshold with more drastic differences for accThreshold values below 0.3 m/s^2. The changes are relatively small for accThreshold greater than 0.3 m/s^2, but most trips in all zones get classified as constant speed state. Based on the results of modeling longitudinal acceleration/deceleration rate for the IHSDM (*8*), an accThreshold of 0.1 m/s^2 was used for the rest of this paper.

Finally, the sensitivity analysis of comfort threshold to minSSWin was examined by comparing the mean and standard deviation values using pvalThreshold of 0.05, accThreshold of 0.1 m/s^2, and five values of minSSWin ranging from 0.1 to 0.5. The values of standard deviation were not affected by the change of minSSWin, with a maximum change in standard deviation value on any curve less than 0.02 g. Figure 2*d* shows that, as expected, the mean comfort threshold on each curve increases with the increase of minSSWin. However, the rate of increase decreases considerably for minSSWin of 0.3 and higher. It was judged that a minimum size of sliding window of 0.3 will provide a good compromise between the advantages of larger and smaller window sizes and was used in the rest of this paper.

Driver Behavior on Horizontal Curves

The procedure and Matlab code developed in this paper were used to process the NDS dataset and study driver behavior. Based on the sensitivity analysis results, the analysis used pvalThreshold of 0.05, accThreshold of 0.1 m/s^2, and minSSWin of 0.3. Each curve was divided into three zones to determine the frequencies of speed state and occurrences of $V_{crv-min}$ and $V_{crv-max}$ in each zone.

Minimum and Maximum Curve Speeds

Figure 3 shows the percentage of trips with $V_{crv-min}$ and $V_{crv-max}$ in each zone of horizontal curves categorized according to terrain (rolling and mountainous) and turning direction (left and right). The figure shows that $V_{crv-min}$ and $V_{crv-max}$ frequencies are almost the same for left and right turn curves. The values of $V_{crv-min}$ and $V_{crv-max}$ were compared for the 12 curves in rolling terrain where both directions were included in the dataset using analysis of variance (ANOVA), and the difference between $V_{crv-min}$ or $V_{crv-max}$ on left and right turn curves was not significant at 5% level of significance for most curves (10 out of 12). On the other hand, a paired *t*-test of hypothesis showed that $V_{crv-min}$ and $V_{crv-max}$ are statistically different at 5% level of significance on 81% of all curves.

Figure 3 also shows that the first and last one-third of the curve (zones 1 and 3) have more occurrences of $V_{crv-max}$ than the middle third (zone 2). This indicates that drivers often do not complete their deceleration on the approach tangent and enter curves with a relatively high speed (causing high $V_{crv-max}$ frequency in zone 1), then decelerate on the curve, and finally accelerate before the end of curve (causing another high $V_{crv-max}$ frequency in zone 3). Thus, modeling speed profile on horizontal curves should account for deceleration at the beginning of curve and acceleration at the end of curve rather than the common assumption of constant speed along the entire curve.

Longitudinal Acceleration

Longitudinal deceleration is an important driving behavior parameter that captures driver response to a stimulus ahead. Larger longitudinal decelerations can indicate delayed driver responses due to reasons such as conspicuous stimuli, unexpected events, inappropriate speed adaptation, and/or driver inattention. As drivers approach a horizontal curve, they tend to decelerate by releasing the throttle paddle or applying the brakes. Deceleration values differ from one driver to another on the same curve depending on factors related to driver characteristics and for the same driver on different curves depending on factors related to terrain, road classification, and general speed environment. Longitudinal deceleration also reflects the longitudinal friction demand, which affects the available lateral friction.

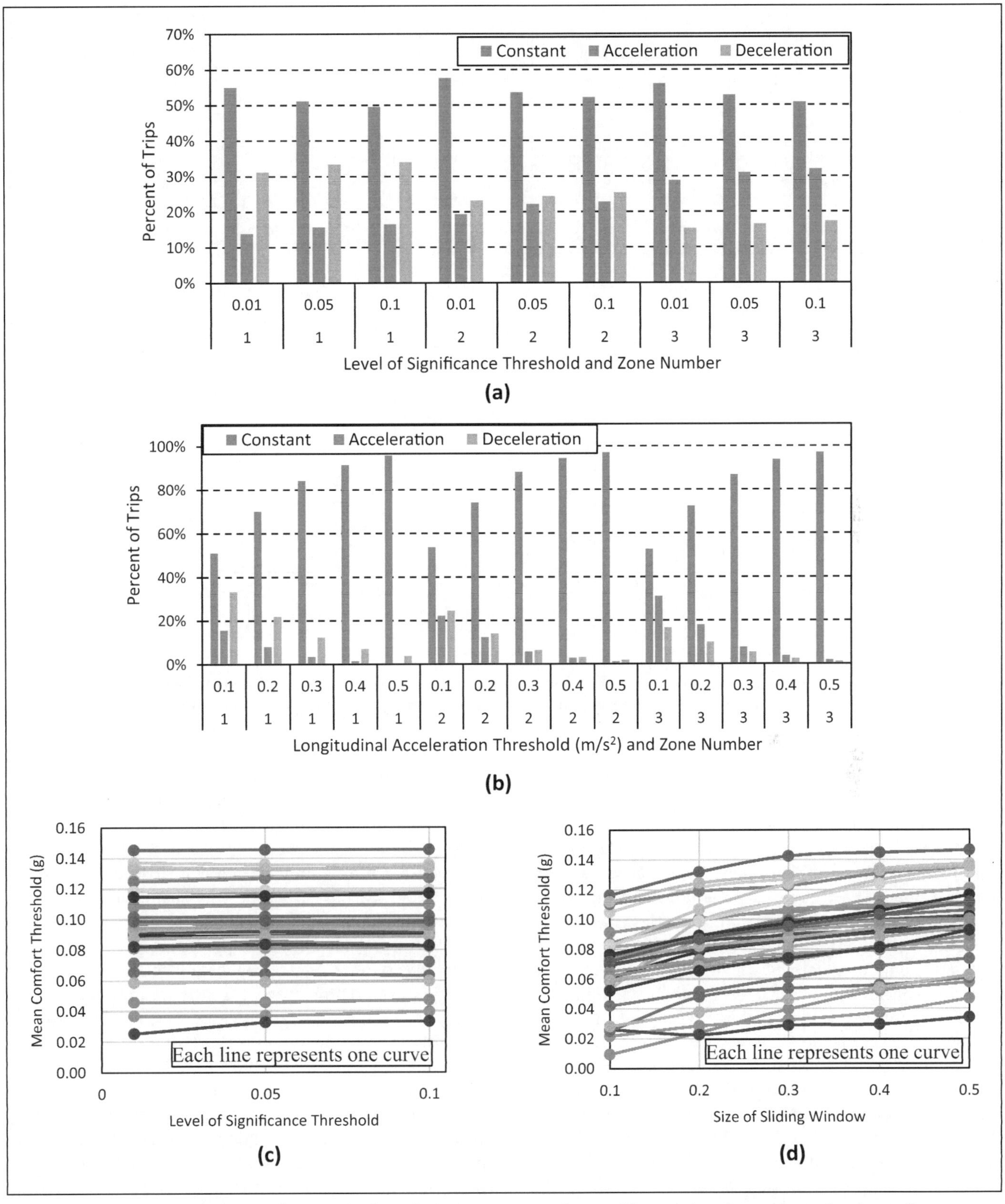

Figure 2. Examples of sensitivity analysis results for rolling terrain curves: change of frequencies of general speed state on all curves with (*a*) pvalThreshold and (*b*) accThreshold; change of mean comfort threshold with (*c*) pvalThreshold and (*d*) minSSWin.

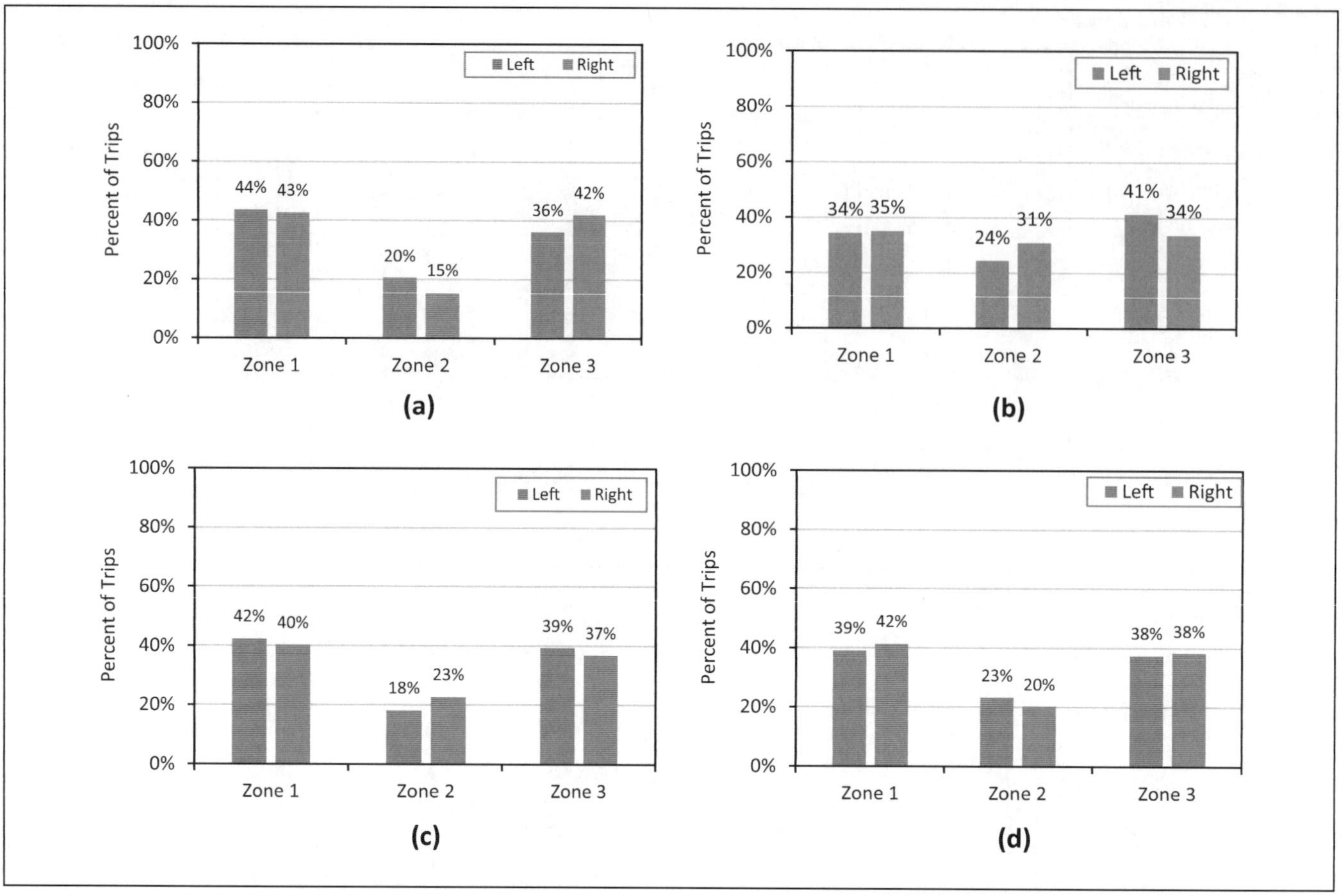

Figure 3. Frequencies of maximum and minimum occurrences in curve zones, showing (*a*) maximum speeds in rolling terrain, (*b*) minimum speeds in rolling terrain, (*c*) maximum speeds in mountainous terrain, and (*d*) minimum speeds in mountainous terrain.

The developed procedure was used to evaluate vehicles' speed state by dividing each curve into three zones of equal length representing three different stages from the points of view of driver perception, curve geometry awareness, and driving adjustment. As explained earlier, the same number of zones was used for all curves to allow aggregation and comparison of the data. As shown in Figure 4, the frequencies of constant speed state are relatively high in all zones. In addition, the constant speed frequencies are higher in mountainous terrain than in rolling terrain and are higher on upgrade than on downgrade. Regardless of the terrain or grade, the highest deceleration and acceleration frequencies are observed in zones 1 and 3, respectively. Finally, the deceleration frequencies are higher in all zones of downgrade than the corresponding zones of upgrade. This can be the result of entering the curves on downgrade with higher speeds.

The same approach was also used to evaluate the vehicles' speed states on the approach tangents. Each tangent was 200 m long divided into three zones with zone 3 ending at BC. As shown in Figure 5*a* for rolling terrain, all frequencies were almost the same at all three zones. Approximately one third of drivers decelerated at each zone on the approach tangent, which means that at each zone about two thirds of the drivers were accelerating or maintaining constant speed. A similar trend was observed for mountainous terrain but the frequency of deceleration trips in each zone was higher on downgrade and lower on upgrade.

Figure 5 further examines the effect of curve radius on the speed state on approach tangents in rolling terrain. First, the figure shows that the frequency of decelerating drivers generally increases as the curve radius decreases. The same trend was obvious on the downgrade but less obvious on the upgrade of mountainous terrain. Some exceptions can exist (e.g., curve radius of 174 m on rolling terrain) which can be related to visual perception of the curve or adopting higher deceleration rates. It can also be noted that the curves with the largest radius had an increase in the frequencies of decelerating vehicles, which can also be related to the adopted deceleration rate. It is therefore recommended to expand the analysis by investigating the relationship between curve radius and deceleration rate.

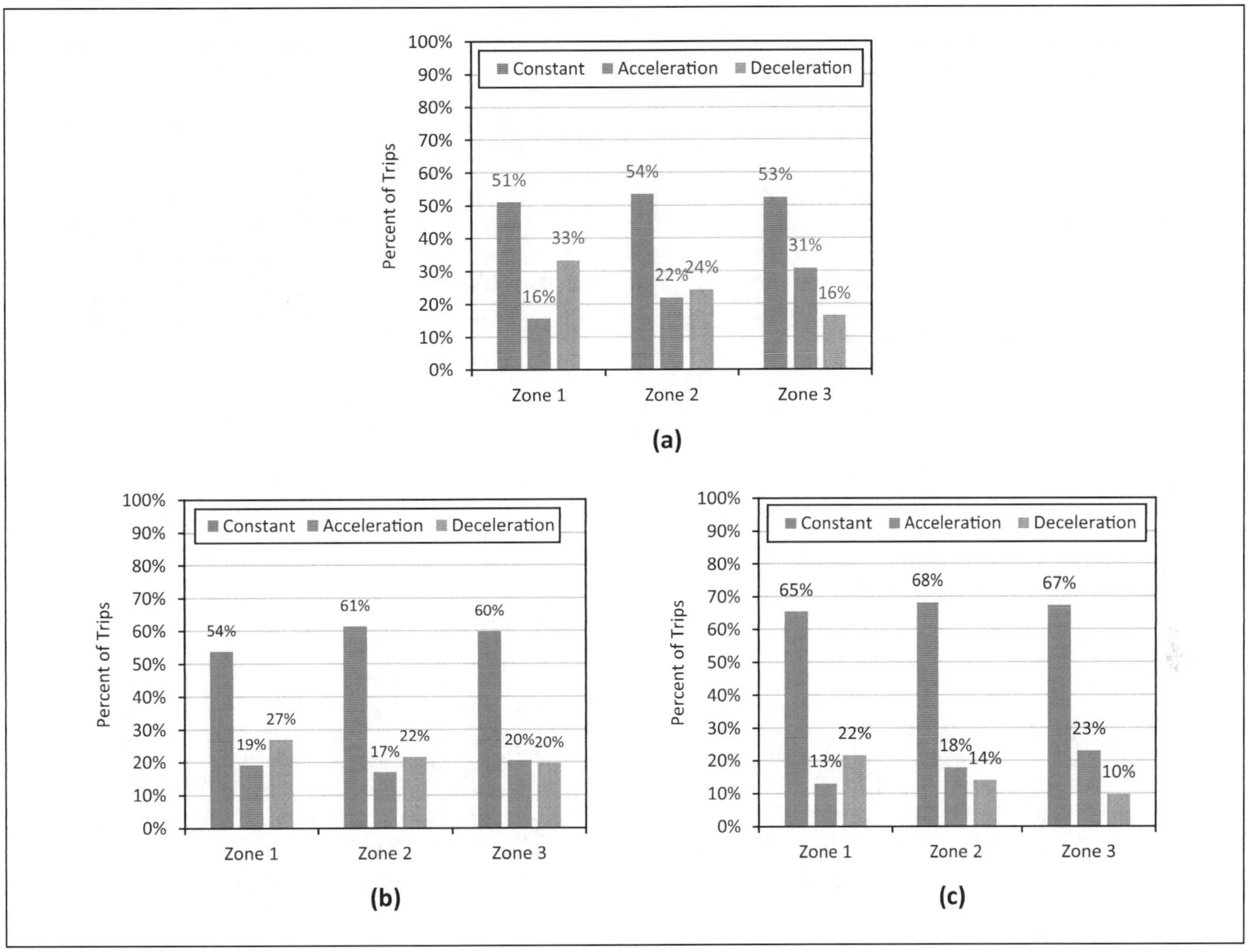

Figure 4. Frequency of general speed state on curve zones for (a) rolling terrain, (b) mountainous terrain downgrade, and (c) mountainous terrain upgrade.

Driver Comfort Threshold

In determining driver comfort threshold, only trips in free-flow conditions should be considered to exclude trips where the lateral forces may be below the comfort threshold due to driving at the speed of traffic rather than the driver's preferred speed. Typically, vehicles are considered in free-flow condition when having a headway (h) of 5 seconds or more (*13*). However, the NDS dataset was used to examine the relationship between headway and speed, and no specific trend was found for the mean or 85th percentile of speed on curve versus headway threshold ($h \geq 1$, $h \geq 2$, ..., $h \geq 5$). On the other hand, the 85th percentile and mean speed on curve for specific headway intervals ($0 \leq h < 1$, $1 \leq h < 2$, ..., $h \geq 5$) showed a trend of increasing speed with increasing headway only for $h < 3$ s. Subsequently, a headway of 3 seconds was used as the threshold for free-flow speed.

The developed procedure was used to determine the comfort threshold for each trip if the trip involves a constant speed state over at least 30% of the curve length. Figure 6 shows that the distribution of comfort thresholds varies with curve radius, and the mean value decreases as the radius increases. For example, the highest frequencies of low comfort threshold (0.00–0.10 g) were experienced on curves with radii greater than 303 m for mountainous terrain and 274 m for rolling terrain. On the other hand, high frequencies of high comfort thresholds were experienced on curves with small radii. Thus, drivers accept higher levels of lateral acceleration, i.e., have higher comfort thresholds, on sharper curves, which confirms the conclusion by Said et al. (*12*) and agrees with horizontal curve design practices.

Finally, Figure 6 shows that the values of comfort thresholds of right turn curves were relatively higher than those of left turns. ANOVA was used to test the null

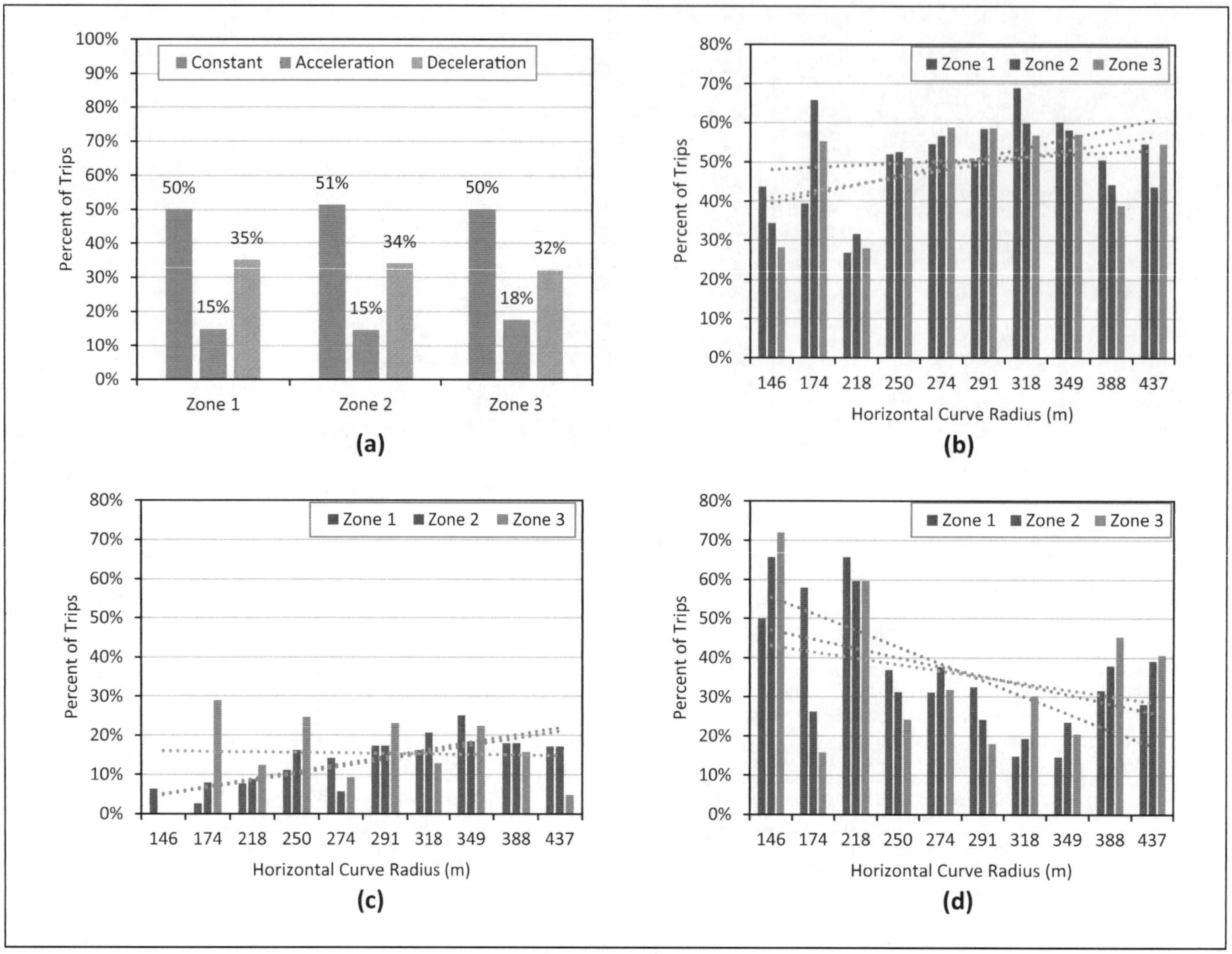

Figure 5. Frequency distribution of general speed state on curve approach tangents in rolling terrain for (*a*) all curve, (*b*) drivers at constant state, (*c*) drivers at acceleration state, and (*d*) drivers at deceleration state.

hypothesis that the mean comfort threshold is equal for left and right turning curves. The test showed that the null hypothesis should be rejected at 5% level of significance on half of the rolling terrain curves considered in both directions (6 out of 12 curves). Although there was no evidence that drivers adopt different speeds on left and right turn curves, it is more likely that they tolerate more lateral forces on right-turn curves compared to left-turn. However, given that the statistical evidence is not very strong, more data should be collected to examine this issue. Furthermore, Figure 6*c* and *d* show lateral friction values in the Green Book (*14*) along with average and 85th percentile of estimated comfort threshold for rolling and mountainous terrains. The Green Book values correspond to the design speeds that would yield the curve radii for 6% superelevation rate. The plots show that the 85th percentile comfort threshold is greater than the Green Book values for curves with sharp radii, indicating that drivers tend to accept higher level of lateral acceleration and thus higher level of risk.

Concluding Remarks

The NDS data have several advantages over traditional data collection approaches utilized in previous research as it is not limited to specific locations or short duration and is free of human or technical errors associated with other data collection techniques. Most importantly, it depicts the naturalistic driving behavior because drivers were not influenced by the presence of observers or measuring devices due to the long duration of study. However, attention should be paid for the quality of recorded data, since some trips were found to lack observations of essential variables, and for the expected reduction in sample size.

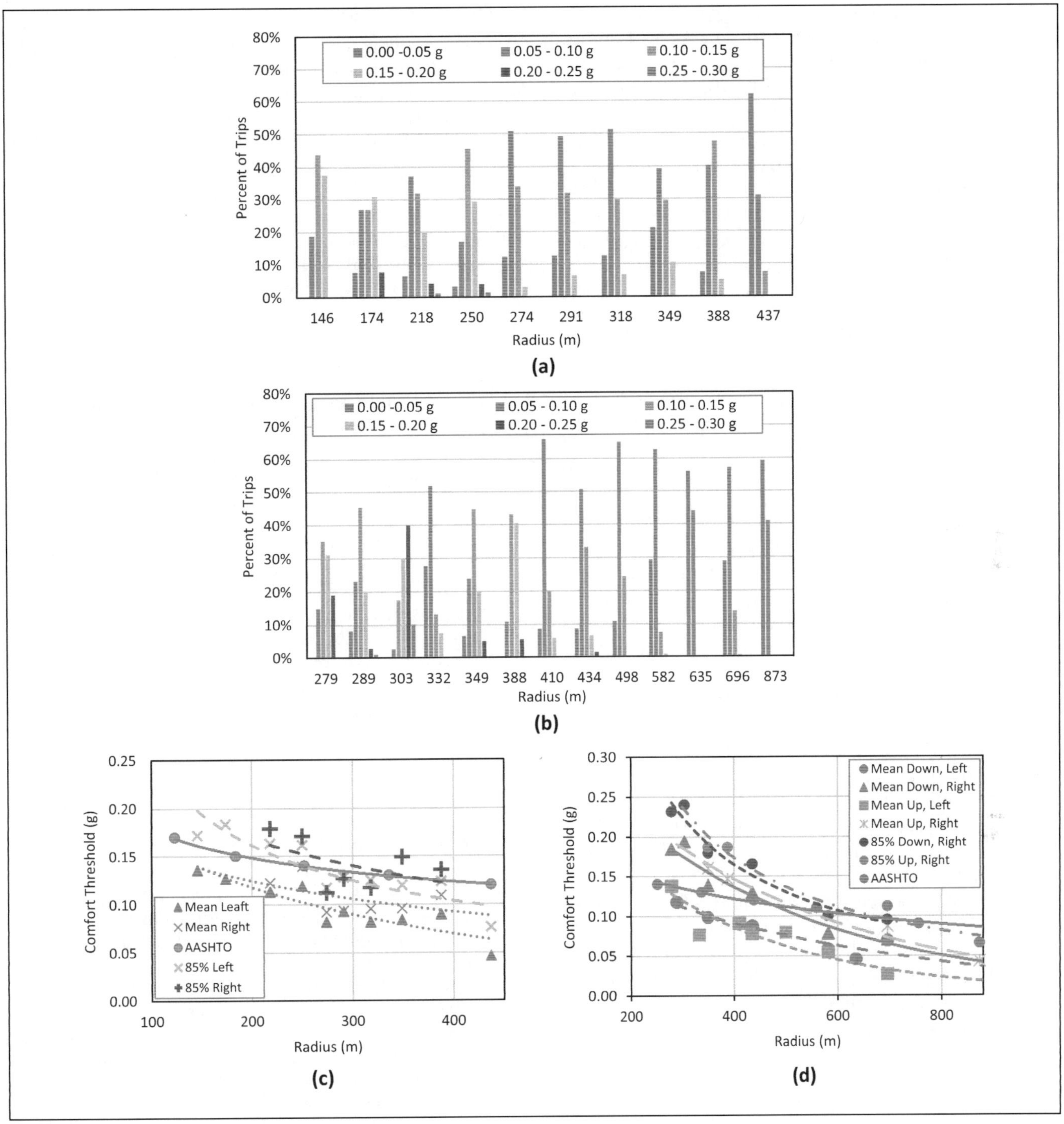

Figure 6. Comfort thresholds values and frequency for different horizontal curve radii: distribution of driver comfort on curves in (*a*) rolling terrain and (*b*) mountainous terrain; effect of radius in (*c*) rolling terrain and (*d*) mountainous terrain.

A methodology and Matlab code were developed to evaluate three driver behavior parameters on horizontal curves: general speed behavior, longitudinal acceleration, and driver comfort threshold. Sensitivity analysis was conducted to examine the effect of user-specified parameters on the results. Applying the developed methodology to a dataset of trips during daytime and favorable weather conditions on 46 horizontal curves in rolling and mountainous terrains on two-lane rural arterial highways, the main findings are as follows:

- The developed methodology can extract key speed and driver comfort measures for developing reliable driver behavior models that are needed for reliability analysis (*5*) and design consistency evaluation (*6*). Results can also be used to examine commonly-accepted assumptions such as the headway for free-flow speed or length of independent tangents.

- Statistical analysis showed no significant difference between speeds on right and left turns of the same curve.
- The general speed trend shows that the assumption of constant speed along horizontal curves is not supported by the data. Rather, speed profile models should account for vehicles not completing the deceleration on the approach tangent, thus decelerating at the beginning of the curve and accelerating at the end.
- The curve middle zone has low frequencies of maximum and minimum curve speeds; thus, further analysis is required to assess the accuracy of modeling curve speed based on measurements at the curve midpoint.
- On all curves, more trips exhibit a constant speed state on all zones (which correspond to subsections) of the curve or approach tangent than any other speed state. On approach tangents, the percentage of trips with a deceleration state increases as the curve radius decreases. Exceptions were observed to this trend suggesting that there is a need to determine an optimal number of zones for speed state analysis and evaluate the actual deceleration rate in each zone.
- Driver comfort thresholds had an inverse relationship with curve radius, and right-turn curves had higher comfort thresholds than left-turn. However, the statistical significance of the difference is not strong, and more analysis is required.
- The 85th percentile values of comfort threshold are greater than those recommended in the Green Book for sharp curves, where drivers might accept higher levels of risk. The extracted data can be used to develop models to predict comfort threshold for horizontal curve design.

Acknowledgments

The authors would like to express their gratitude to Kevin Perkins (WSDOT-Multimodal Planning Division) for providing necessary data and information to conduct the analysis of this research. The authors are highly indebted to Miguel Perez, Christina Witcher, and Whitney Atkins (Virginia Tech Transportation Institute, VTTI) for their help in providing the naturalistic driving behavior data. Financial support by the Natural Sciences and Engineering Council (NSERC) of Canada is gratefully acknowledged.

Author Contributions

The authors confirm contribution to the paper as follows: study conception and design: Bashar Dhahir, Yasser Hassan; data collection: Bashar Dhahir; analysis and interpretation of results: Bashar Dhahir, Yasser Hassan; draft manuscript preparation: Bashar Dhahir, Yasser Hassan. All authors reviewed the results and approved the final version of the manuscript.

References

1. Misaghi, P., and Y. Hassan. Modeling Operating Speed and Speed Differential on Two-Lane Rural Roads. *Journal of Transportation Engineering*, Vol. 131, No. 6, 2005, pp. 408–417.
2. *Special Report 260: Strategic Highway Research: Saving Lives, Reducing Congestion, Improving Quality of Life.* TRB, National Research Council, Washington, D.C., 2001.
3. Himes, S. C., E. T. Donnell, and R. J. Porter. Posted Speed Limit: To Include or Not to Include in Operating Speed Models. *Transportation Research Part A: Policy and Practice*, Vol. 52, 2013, pp. 23–33.
4. Hassan, Y. Highway Design Consistency: Refining the State of Knowledge and Practice. *Transportation Research Record: Journal of the Transportation Research Board*, 2004. 1881: 63–71.
5. Dhahir, B., and Y. Hassan. Reliability-Based Design of Horizontal Curves on Two-Lane Rural Highways. *Transportation Research Record: Journal of the Transportation Research Board*, 2016. 2588: 22–31.
6. Dhahir, B., and Y. Hassan. Relationship between Traffic Safety and Speed Differential on Horizontal Curves Based on Naturalistic Driving Studies. *Proc., 2017 Road Safety & Simulation International Conference*, Paper #58, The Hauge, Netherlands, 2017.
7. Dingus, T., J. M. Hankey, J. F. Antin, S. E. Lee, L. Eichelberger, K. E. Stulce, D. McGraw, M. Perez, and L. Stowe. *SHRP 2 Rep. S2-S06-RW-1: Naturalistic Driving Study: Technical Coordination and Quality Control.* Transportation Research Board of the National Academies, Washington, D.C., 2015.
8. Fitzpatrick, K., L. Elefteriadou, D. W. Harwood, J. M. Collins, J. McFadden, I. B. Anderson, R. A. Krammes, N. Irizarry, K. D. Parma, K. M. Bauer, and K. Passetti. *Speed Prediction for Two-Lane Rural Highways.* Report FHWA-RD-99-171. Federal Highway Administration, Washington, D.C., 1999.
9. Geoscience Australia. *Applying Geoscience to Australia's Most Important Challenges.* http://www.ga.gov.au/scientific-topics/astronomical/astronomical-definitions#heading-2. Accessed February 10, 2017.
10. Ranjan, V. *Flight Dynamics, Simulation, and Control for Rigid and Flexible Aircraft.* CRC Press, Tylor and Francis Group, Boca Raton, Fla., 2015.
11. Milliken, W., and D. Milliken. *Race Car Vehicle Dynamics.* Society of Automative Engineers, Warrendale, Pa., 1995.

12. Said, D., Y. Hassan, and A. O. Abd El Halim. Comfort Thresholds for Horizontal Curve Design. *Canadian Journal of Civil Engineering*, Vol. 36, No. 12, 2009, pp. 1391–1402.
13. Hassan, Y., M. Sarhan, and R. Porter. (ed.). *Transportation Research Circular E-C151: Modeling Operating Speed Synthesis Report*. Transportation Research Board of the National Academies, Washington, D.C., July 2011.
14. American Association of State Highway and Transportation Officials. *A Policy on Geometric Design of Highways and Streets*. American Association of State Highway and Transportation Officials, Washington, D.C., 2011.

The Standing Committee on Operational Effects of Geometrics (AHB65) peer-reviewed this paper (18-02863).

The findings and conclusions of this paper are those of the authors and do not necessarily represent the views of the VTTI, SHRP 2, the Transportation Research Board, or the National Academy of Science.

Article

TRR
JOURNAL OF THE TRANSPORTATION RESEARCH BOARD

Transportation Research Record
2018, Vol. 2672(17) 96–107

DOI: 10.1177/0361198118792754
journals.sagepub.com/home/trr
SAGE

Operational Effects of the Consolidated Intersection Design on Urban and Suburban Arterials

Daniel J. Cook[1]

Abstract

Along urban and suburban arterials, closely-spaced signalized intersections are commonly used to provide access to adjacent commercial developments. Often, these signalized intersections are designed to provide full access to developments on both sides of the arterial and permit through, left-turn, and right-turn movements from every intersection approach. Traffic signal timing is optimized to reduce vehicle delay or provide progression to vehicles on the arterial, or both. However, meeting both of these criteria can be cumbersome, if not impossible, under high-demand situations. This research proposes a new design that consolidates common movements at three consecutive signalized intersections into strategic fixed locations along the arterial. The consolidation of common movements allows the intersections to cycle between only two critical phases, which, in turn, promotes shorter cycle lengths, lower delay, and better progression. This research tested the consolidated intersection concept by modeling a real-world site in microsimulation software and obtaining values for delay and travel time for multiple vehicle paths along the corridor and adjacent commercial developments in both existing and proposed conditions. With the exception of unsignalized right turns at the periphery of the study area, all non-displaced routes showed a reduction in travel time and delay. Additional research is needed to understand how additional travel through the commercial developments adjacent to the arterial may effect travel time and delay. Other expected benefits of the proposed design include a major reduction in conflict points, shorter pedestrian crossing and wait times, and the opportunity to provide pedestrian refuge areas in the median.

Urban and suburban arterials serve dual purposes: offering mobility for traffic traveling on the arterial, and providing access to adjacent properties and other streets. Large commercial centers tend to impose stress on the operation of the arterials that serve them, because the intersections that provide access to the property must maintain a level of service for through traffic on the arterial while also providing drivers (and pedestrians and cyclists) with access to desired destinations. When these commercial centers flank both sides of the arterial and multiple (sometimes closely-spaced) signalized intersections are used to provide access to these developments, it becomes very difficult to provide adequate levels of service to the side-street movements while at the same time providing bidirectional progression for the through traffic on the arterial. As these competing demands increase, cycle lengths and green splits are increased, often resulting in increased signal delay and longer queues.

Equation 19-16 of the *Highway Capacity Manual* (HCM) (*1*) defines capacity of a lane group for which there are no permitted left-turn movements as the product of the number of lanes in the lane group, the saturation flow rate of the lane group per lane, and the green split. This equation is shown below:

$$c = Ns\frac{g}{C} \tag{1}$$

where

c = capacity (veh/h)
N = number of lanes in lane group
s = saturation flow rate (veh/lane/h)
g = effective green time (s)
C = cycle length (s)

According to this equation, increasing the number of lanes in the lane group (N) increases capacity. However, adding lanes can be a very expensive solution for increasing capacity at signalized intersections along an arterial

[1]MRIGlobal, Kansas City, MO

Corresponding Author:
Address correspondence to Daniel J. Cook: dcook@mriglobal.org

with commercial developments on both sides. Increasing the green split (g/C) can also increase capacity. This can be accomplished by reducing the number of critical phases at an intersection.

There are several examples of alternative intersection designs that reduce the number of critical phases, accomplishing the goal of increasing capacity at the intersection without the need for additional through lanes (although additional auxiliary lanes are sometimes needed for channelization). These intersection types include:

- Restricted crossing U-turn
- Median U-turn
- Displaced left-turn (also called continuous flow)

These intersection types have been heavily documented in the *HCM* (*1*) and other research papers (*2–4*).

The *HCM* states that alternative intersection designs have significantly reduced travel times and delay, reduced the number of conflict points between vehicles, and can often be implemented with minimal disruptions to existing right-of-way.

The *HCM* also defines "distributed intersections" as a group of two or more intersections that operate interdependently by virtue of close spacing and displaced or distributed traffic movements. The level of service for a route in a distributed intersection is based on the travel time experienced by the driver, which the *HCM* refers to as experienced travel time (*ETT*). *ETT* is the sum of all control delay experienced by the driver traversing the distributed intersection and all the travel time due to extra distance required to be traveled in the distributed intersection.

Adjacent intersections that serve the same commercial developments along an arterial have redundant movements. For example, a driver traveling westbound on the arterial who wants to turn left into the commercial development on the south side of the arterial can use any of the intersections along the arterial that provide access to this development. Similarly, a driver wanting to cross from the development on the north side of the arterial to the development on the south side has several route choices. This study proposes a new distributed intersection design for the arterial that consolidates redundant vehicle paths at three adjacent intersections by limiting turning and through movements at each intersection. Access to or from the commercial developments is provided, generally, by a single route to or from the arterial or between developments.

The goal of this study was to create a new arterial design that improved traffic operations for all movements on an arterial that is adjacent to commercial development on both sides of the roadway. The base condition is a segment of urban or suburban arterial with three consecutive fully accessible intersections, where there is commercial development on both sides of the arterial. The three intersections provide access to these commercial developments but do not provide access to other developments or neighborhoods. Figure 1*a* shows a basic plan view of the base condition with all permissible movements shown in red.

The proposed design, shown in Figure 1*b*, consolidates common movements together to eliminate many of the redundant turning movements. This design requires all traffic within the commercial development to access each of the three intersections via roadways internal to the commercial development, such as a frontage road or a ring road. Without proper internal circulation in the commercial development, the proposed design will not function properly because displaced routes will not be able to access the proper intersection to make their turn onto the arterial.

This study uses the term "consolidated intersection design" to refer to the proposed solution shown in Figure 1*b*. The singular form of intersection is used in this phrase because all three individual intersections from the base condition now function together as a distributed intersection. Throughout this report, the term "consolidated intersection" refers to the entire arterial and its three access points to adjacent commercial developments.

The consolidated intersection is designed to utilize existing right-of-way. Each of the traffic signals within the consolidated intersection operates in a two-critical-phased operation. All arterial traffic turning left must use the first access point they approach. All traffic originating from the commercial development must use the last access point available to make a right turn onto the arterial, and the center access point to make a left turn onto the arterial. To move from the development on one side of the arterial to the development on the other side, drivers must use the center access point followed by an immediate right-turn to go across the arterial to the other commercial development. The dashed lines shown in Figure 1*b* represent movements that could potentially bypass traffic control, because they have no conflicting crossing movements.

All intersections in the consolidated intersection design can be operated with just a two-critical-phased signal operation. This fact lends itself to lower optimal cycle lengths than the cycle lengths that would be required to operate the intersections in the base condition. Lower cycle lengths have a high potential for lower delay. Lower delay means lower travel times throughout the consolidated intersection. The two-critical-phased signal operation also allows for more flexibility in optimizing bidirectional progression on the arterial.

Methodology

Microsimulation was used to test the hypothesized operational improvements of the consolidated intersection

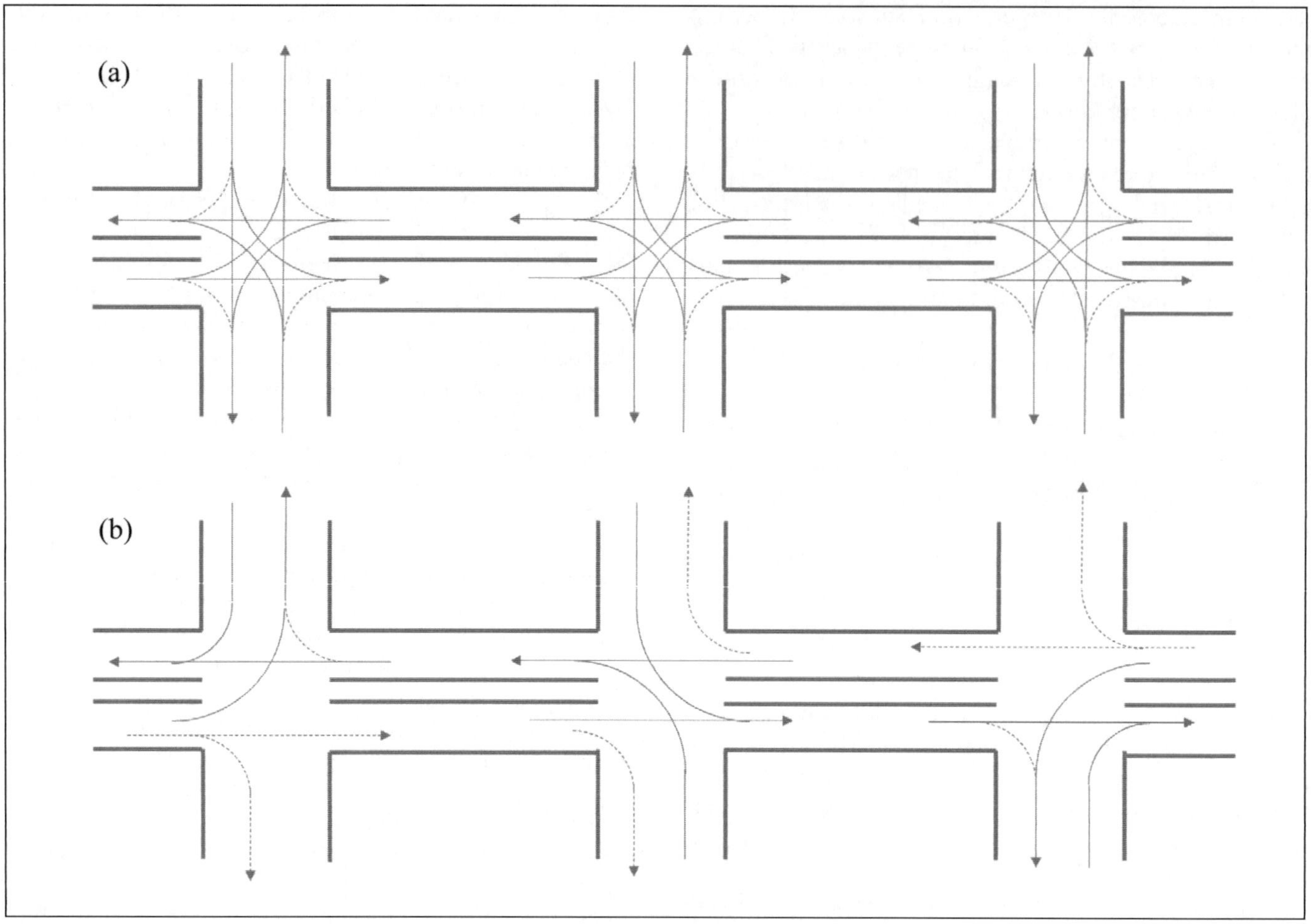

Figure 1. (*a*) Base condition and (*b*) proposed design permissible movements.

design. Vissim 8 was used as the microsimulation platform. A real-world site was selected to represent a baseline condition against which to test the proposed consolidated intersection. A calibrated microsimulation model was built for the base condition, and then a microsimulation model of the proposed design was built. The operations of the internal circulation routes were not included in the simulation. The operational characteristics of each condition were gathered from the microsimulation and compared with a statistical analysis.

As discussed previously, the consolidated intersection is intended to improve operations along urban and suburban arterials in which there are three closely-spaced intersections, commercial development on both sides, and internal circulation within the developments on each side of the arterial. A baseline location was sought with these parameters in mind. Also, a site in the Kansas City metropolitan area was desired to allow the research team access for site visits.

39th Street in Independence, Missouri, was selected as the arterial to serve as the base condition in the study. There are three intersections along 39th Street included in the analysis area. For simplicity, these intersections will be referred to as Western, Center, and Eastern intersections throughout the rest of the paper, correlating with the intersections' relative geographical position along 39th Street. Figure 2 shows an aerial view of the study area from Google Earth.

The Eastern intersection is an unsignalized intersection with stop control on the side streets. Although it would have been desirable to have all three intersections signalized, the unsignalized intersection was included in the model to remain as true as possible to existing conditions. A large enclosed mall with several big-box retail stores is located on the south side of the arterial. This mall is served by a ring road that distributes mall traffic to the three arterial access points. A strip mall and several restaurants and retail stores are located on the north side of the arterial. Note that modifications would be required to adequately accommodate displaced routes on the north side of the arterial, including a public street connection between points 8 and 7.

Turning movement counts, origin–destination (O-D) data, mainline travel-time, and other observations were

Figure 2. Google Earth aerial view of study area.

collected for the study area. The purpose of this research was to test a new intersection concept using realistic data that represents real-world conditions, rather than to design a solution for a specific location. The data collection performed in this study was used to provide a realistic baseline condition rather than to calibrate the model to exactly match existing conditions. Therefore, the data collection performed for this study was not as detailed as would be needed to design a solution for this specific location.

Turning movement counts were collected on a weekday afternoon during the week before Christmas, when shopping traffic was high, but regular weekday traffic was still present as well. This combination of turning traffic and through traffic presented the optimum conflicting volumes to present a good test for the consolidated intersection design. Heavy vehicles were classified separately during the data collection, however very few trucks were present. For the sake of the simulation analysis, 1% of the vehicle fleet was assumed to be heavy vehicles.

There are eight points in which a vehicle can enter the study area and eight points where a vehicle can exit the study area. Figure 2 shows how these access points are labeled in this report. For example, an eastbound vehicle on 39th Street enters the study area at Point 1. If that vehicle turns left at the Center intersection, the O-D route for this vehicle is 1 to 7. Initial observations of the study area showed that no traffic entered the arterial to later turn back into the same commercial development downstream (e.g., O-D route 8 to 7).

To collect O-D data, video cameras were used to record vehicles traveling within and through the study area. The video was later reduced to produce O-D percentages for each approach. These O-D percentages were multiplied by the origin demand volume to determine the volume for each route. Figure 3 shows the O-D routes' volumes in the study area.

Vissim 8 was used to build a microsimulation model of the existing baseline condition and a model of the consolidated intersection design. The base condition model approximates field conditions, using turning movement counts and O-D data collected in the field. The three intersections modeled in the base simulation model are shown in Figure 4. It is important to note that nonmotorists were not modeled in the simulation even though pedestrians were observed sparingly on the corridor. The scope of this study focused on the operational effects of the proposed intersection design on motor vehicles. See the Discussion section for more details on nonmotorists.

Field timing plans were not used in the simulation model. The signals at the Western and Center intersections were optimized with in-house software using the collected turning movement counts in order to provide optimum progression for both east- and westbound directions as well as to optimize level of service for side-street approaches. The resulting phasing diagrams are shown in Figure 5.

The main goal of the proposed design was to utilize existing right-of-way without the need for additional space. This goal is quite achievable given that several intersection turn lanes in the study are removed, freeing

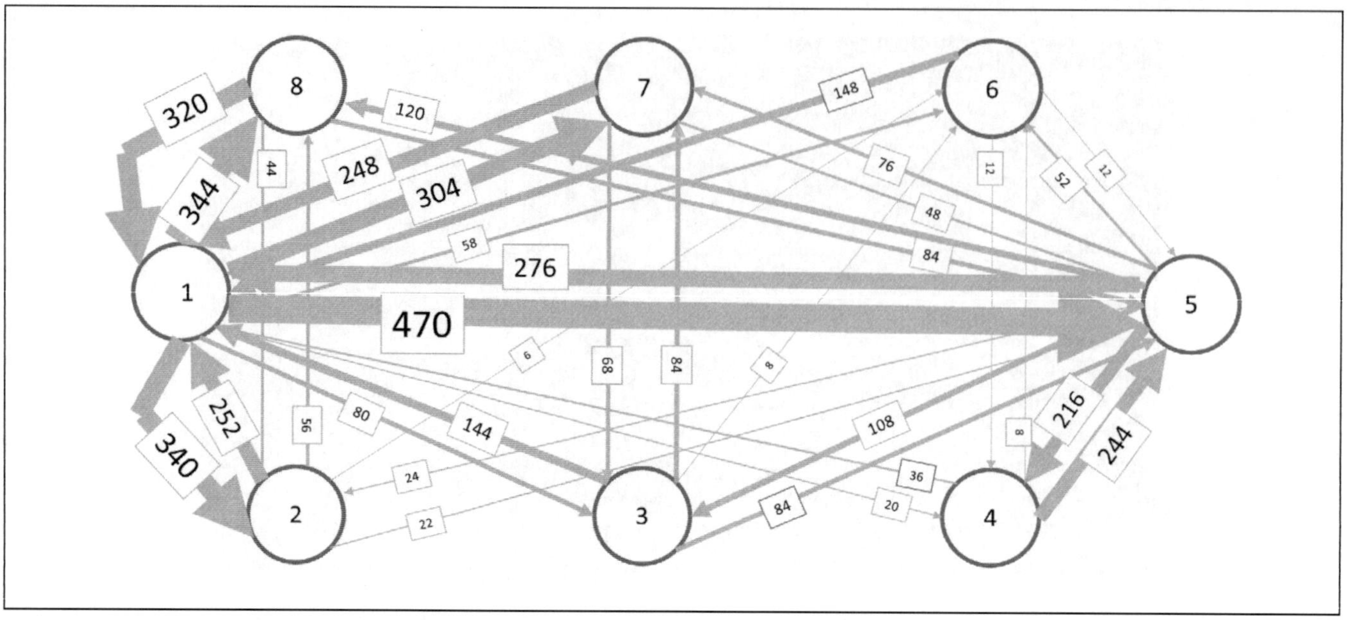

Figure 3. O-D route design hour volumes.

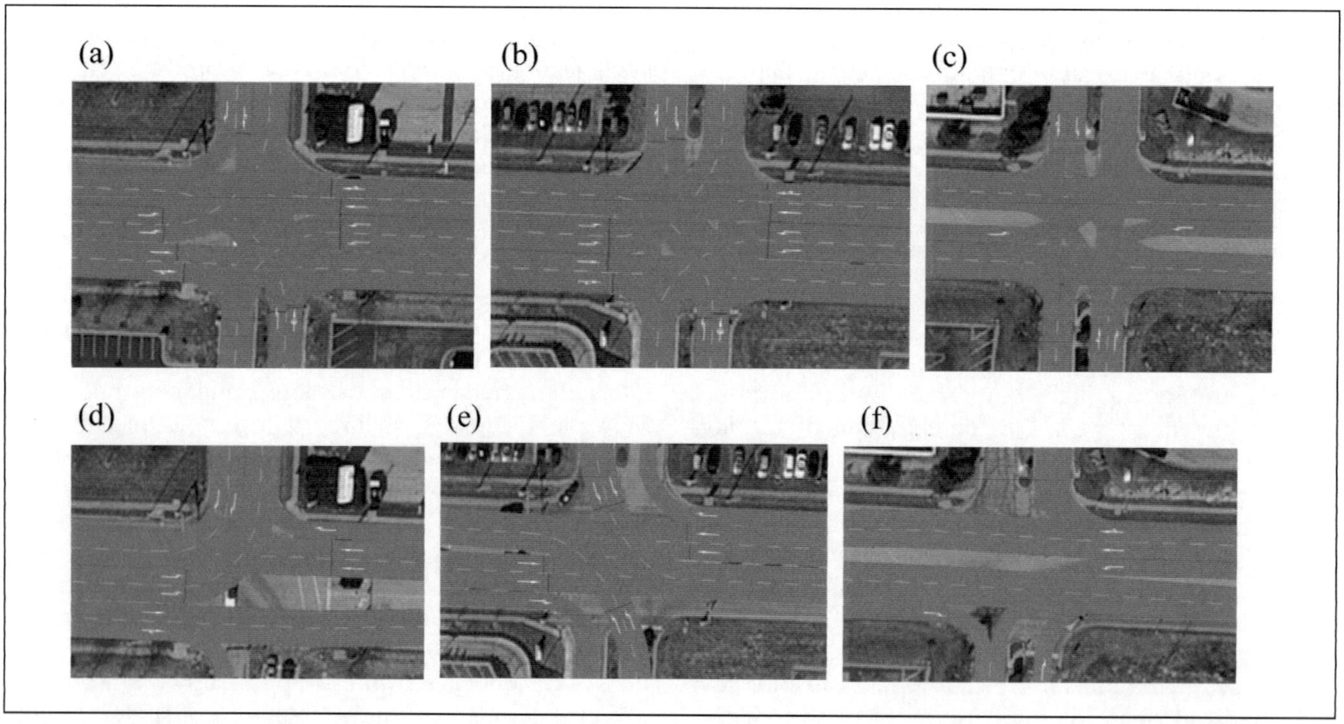

Figure 4. Intersection configurations in simulation models. Base model: (*a*) Western; (*b*) Center; (*c*) Eastern. Consolidated design model: (*d*) Western; (*e*) Center; (*f*) Eastern.

space within the right-of-way to be used for other purposes.

The outer intersections of the consolidated intersection design only allow left turns from the mainline direction entering the study area. That is, eastbound traffic can turn left from the arterial into the development on the north side of the arterial only at the Western intersection (and vice versa). The left-turn movement's complementary right-turn is the only movement permitted from the side street. The eastbound through and right-turn movements of vehicles entering the study area on the arterial are essentially unimpeded, except when

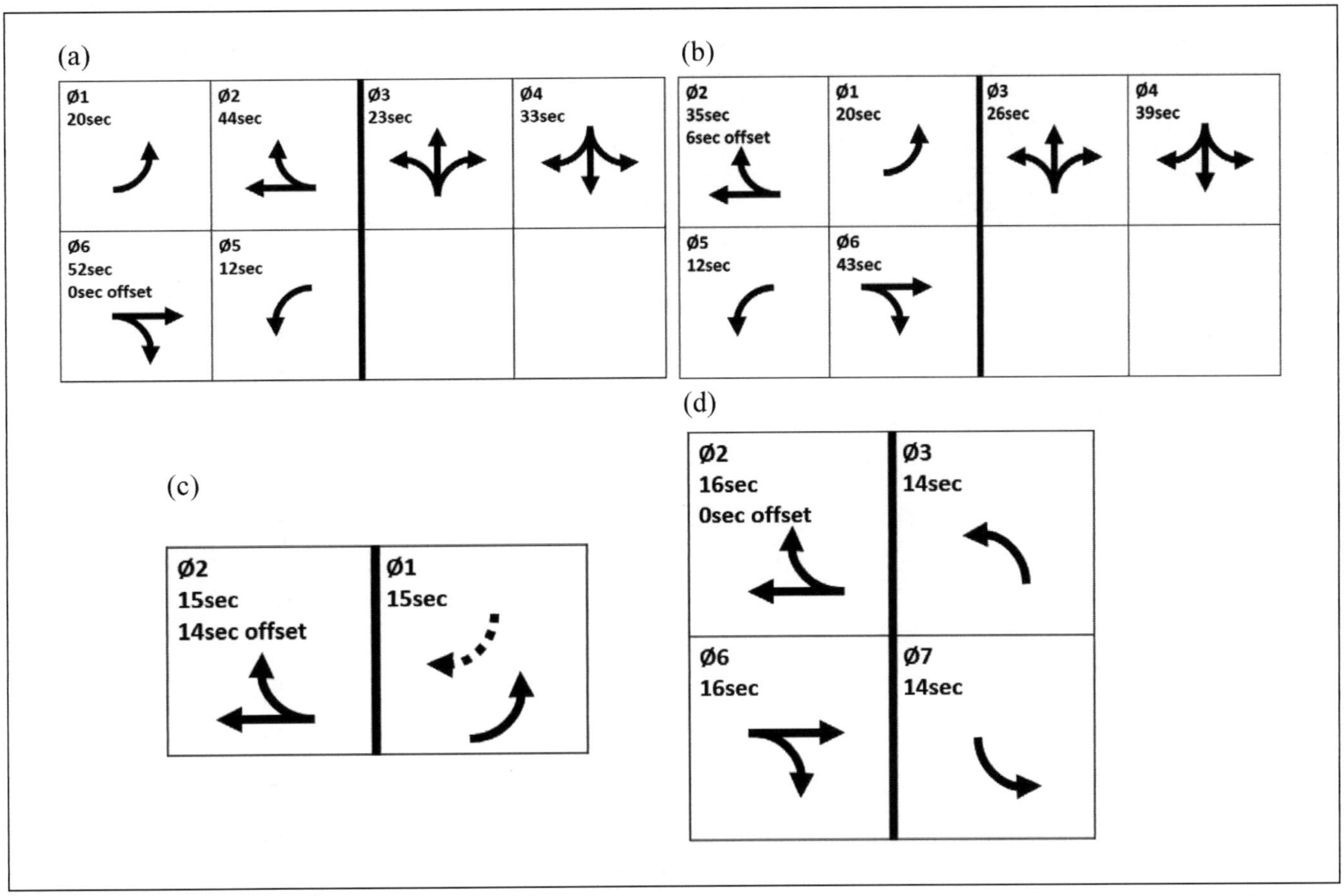

Figure 5. Optimized signal timing plans used in the simulation models: (*a*) Western intersection, before; (*b*) Center intersection, before; (*c*) Western intersection, new; (*d*) Center intersection, new.

pedestrians are present (nonmotorists are discussed later in the report). Figure 4*d* shows the Western intersection configuration in detail.

Left-turn movements are the only permissible movements on the side streets at the Center intersection. Left turns from the arterial are prohibited, and right turns from the arterial are free-flow. All traffic from the side street traveling to the commercial development on the other side of the arterial must turn left at the Center intersection, then make an immediate right-turn at the downstream intersection. Figure 4*e* shows a detailed layout of the Center intersection. Note that because left turns are no longer permitted from the arterial at this intersection, the opposing through lanes can be moved closer together to allow for the accommodation of right-turn lanes on the arterial.

The access permitted at the Eastern intersection is identical to that of the Western intersection. However, the operation is quite different. In the base condition, there was only stop control on the minor approaches. There was concern about whether the westbound left-turn could handle the increased volume due to the consolidation of all westbound left-turns in the study area. However, since the only conflicting movements are the opposing through movements, it was deemed unnecessary to either provide signal control or provide an extra left-turn lane to accommodate the additional volume. Figure 4*f* shows a detailed layout of the Eastern intersection.

Whether one, two, or three intersections in the consolidated intersection are signal-controlled, optimizing the signal timings to provide bidirectional progression and minimal delays is quite simple. The core philosophy of the consolidated intersection is to only have a two-critical-phased signal operation at each signalized intersection. This allows for lower cycle lengths, thus lower delays and queue lengths (in most situations). Furthermore, the arterial through movements only proceed through two signals. As discussed previously, the through movement at the first intersection entering the study area is unimpeded. As the through movements in both directions of the arterial receive green at the same time at the Center intersection, providing downstream greens for the platoons in each direction is simple. The

Table 1. Change in Travel Time for Common Routes

Base O-D route	CID O-D route	Travel time change[a] (s/veh)	Percent change[a]	*P*-value
4 to 6	3 to 8	22.504	70.8%	<.01
6 to 4	7 to 4	19.036	69.8%	<.01
4 to 1	3 to 1	−37.0216	−40.0%	<.01
4 to 5	4 to 5	−4.9674	−20.5%	<.01
6 to 5	7 to 5	14.348	53.9%	<.01
6 to 1	8 to 1	−68.9153	−77.7%	<.01
3 to 5	4 to 5	−52.377	−73.2%	<.01
3 to 6	4 to 5	−23.3242	−30.0%	<.01
3 to 7	3 to 8	−22.006	−28.8%	<.01
3 to 1	3 to 1	−74.3411	−57.2%	<.01
7 to 5	7 to 5	−35.2556	−46.3%	<.01
7 to 4	7 to 4	−33.5287	−42.0%	<.01
7 to 3	7 to 4	−19.8448	−30.0%	<.01
7 to 1	8 to 1	−59.6483	−75.1%	<.01
2 to 5	4 to 5	−107.582	−84.8%	<.01
2 to 6	3 to 8	−73.139	−57.4%	<.01
2 to 8	3 to 8	−23.88	−30.5%	<.01
2 to 1	3 to 1	−18.2379	−24.7%	<.01
8 to 5	7 to 5	−74.7225	−64.6%	<.01
8 to 2	7 to 4	−18.6585	−28.7%	<.01
8 to 1	8 to 1	−5.5745	−22.0%	<.01
5 to 4	5 to 4	1.7571	8.4%	<.01
5 to 6	5 to 6	0.059	0.3%	1
5 to 3	5 to 4	−69.174	−75.3%	<.01
5 to 7	5 to 7	−23.1558	−40.7%	<.01
5 to 2	5 to 4	−89.5736	−79.8%	<.01
5 to 8	5 to 8	−28.1174	−33.4%	<.01
5 to 1	5 to 1	−27.4598	−33.6%	<.01
1 to 2	1 to 2	−14.3138	−42.0%	<.01
1 to 8	1 to 8	−44.6779	−64.1%	<.01
1 to 3	1 to 3	−23.3194	−41.0%	<.01
1 to 7	1 to 8	−58.8182	−70.2%	<.01
1 to 4	1 to 4	−22.4626	−29.4%	<.01
1 to 6	1 to 8	−54.2284	−68.5%	<.01
1 to 5	1 to 5	−22.8728	−31.5%	<.01

Note: CID = consolidated intersection design.
[a]Positive value represents increase in travel time with consolidated intersection compared with existing design.

offset of the downstream greens would be the travel time of the platoons from the Center intersection to the downstream signals.

The signal timings were optimized with in-house software using the updated turning movement counts in order to provide optimum progression for the westbound direction (eastbound traffic only uses one signal) as well as to optimize level of service for side-street approaches. The resulting phasing diagrams are shown in Figure 5*c* and 5*d*. The resulting cycle length is very short relative to the base condition's cycle length.

Guidance on calibration and validation was obtained through two sources: *Microscopic Simulation Model Calibration and Validation Handbook* (*5*), and *Protocol for Vissim Simulation* (*6*).

The base design simulation model was run four times with different random seeds to collect a baseline to compare travel time data on the mainline with travel time data collected from 39th Street. The travel time was collected from the beginning of the first intersection traversed to the beginning of the last intersection traversed. The averages of the simulation and real-world data set were compared to see if there was any statistical difference. The eastbound average travel time from the simulation was 41.3 s and 42.0 s from the field data. A *t*-test showed no statistical difference between the two mean travel times. The westbound average travel time from the simulation was 66.0 s and 43.3 s from the field data. A *t*-test showed these averages to be significantly different. Westbound vehicles in the field arrive in a platoon formed by upstream signalized intersections, whereas in the simulation westbound vehicles arrive randomly. This fact explains the differences in the average westbound travel times between the simulation and field data. As this study is analyzing the operational effectiveness of the consolidated intersection using real-world data, and

Table 2. Change in Delay for Common Routes

Base O-D route	CID O-D route	Delay change[a] (s/veh)	Percent change[a]	*P*-value
4 to 6	3 to 8	4.1393	26.7%	<.01
6 to 4	7 to 4	–0.0943	–0.8%	1
4 to 1	3 to 1	–25.9047	–53.9%	<.01
4 to 5	4 to 5	–2.9997	–48.2%	<.01
6 to 5	7 to 5	–1.2765	–11.3%	0.51
6 to 1	8 to 1	–38.7196	–83.6%	<.01
3 to 5	4 to 5	–36.4756	–91.9%	<.01
3 to 6	4 to 5	–23.5676	–54.6%	<.01
3 to 7	3 to 8	–35.946	–64.7%	<.01
3 to 1	3 to 1	–73.9769	–77.0%	<.01
7 to 5	7 to 5	–35.5504	–78.1%	<.01
7 to 4	7 to 4	–34.4766	–74.7%	<.01
7 to 3	7 to 4	–37.5546	–76.3%	<.01
7 to 1	8 to 1	–43.9213	–85.3%	<.01
2 to 5	4 to 5	–77.2858	–96.0%	<.01
2 to 6	3 to 8	–59.2583	–75.1%	<.01
2 to 8	3 to 8	–38.3086	–66.1%	<.01
2 to 1	3 to 1	–31.2778	–58.6%	<.01
8 to 5	7 to 5	–63.8494	–86.5%	<.01
8 to 2	7 to 4	–41.575	–78.1%	<.01
8 to 1	8 to 1	–5.5461	–42.2%	<.01
5 to 4	5 to 4	0.9722	20.0%	<.01
5 to 6	5 to 6	–0.015	–2.3%	1
5 to 3	5 to 4	–52.5833	–90.0%	<.01
5 to 7	5 to 7	–23.1995	–96.0%	<.01
5 to 2	5 to 4	–59.7987	–91.1%	<.01
5 to 8	5 to 8	–27.8717	–71.6%	<.01
5 to 1	5 to 1	–27.412	–72.3%	<.01
1 to 2	1 to 2	–14.3009	–88.4%	<.01
1 to 8	1 to 8	–44.8443	–82.8%	<.01
1 to 3	1 to 3	–23.8064	–90.6%	<.01
1 to 7	1 to 8	–43.9325	–82.5%	<.01
1 to 4	1 to 4	–23.0978	–74.1%	<.01
1 to 6	1 to 8	–25.1498	–73.0%	<.01
1 to 5	1 to 5	–22.8481	–75.3%	<.01

Note: CID = consolidated intersection design.
[a]Positive value represents increase in delay with consolidated intersection compared with existing design.

not a design solution for a particular location, it was deemed unnecessary to calibrate the base simulation model any further for arterial traffic.

While collecting data, it was observed that vehicles turning right at the Eastern intersection minor approaches did not come to a complete stop if no conflicting vehicles were present. However, all vehicles in the model turning right at the stop signs came to a complete stop every time. To adjust for this, the stop control was removed for right-turning vehicles in the simulation, and replaced with yield control. A reduced speed zone was placed on the right turn to force vehicles to slow to a speed no higher than 5 mph to replicate a vehicle rolling through a stop sign. If there were oncoming conflicting vehicles in the simulation, the right-turning vehicle would indeed stop before proceeding; otherwise, the vehicle would slowly roll through the stop sign.

Data Analysis

Each simulation model was run 100 times with different random seeds. Each model ran for 4,500 s, but data was collected only for the time period 900 to 4,500 s (1 h). Travel time and delay data were collected for each vehicle and grouped by each route. Travel time for each vehicle started at a point 240 ft upstream of the first intersection traversed on the route, and the travel time ended for each vehicle at a point 120 ft downstream of the last intersection traversed.

There were 35 unique O-D routes in the base model for which there was traffic volume. Several of these 35 O-D routes were consolidated into common O-D routes in the proposed design, which produces 16 unique O-D routes in the consolidated intersection model. When comparing travel time and delay between the two models, it is important to compare the routes that a driver would have

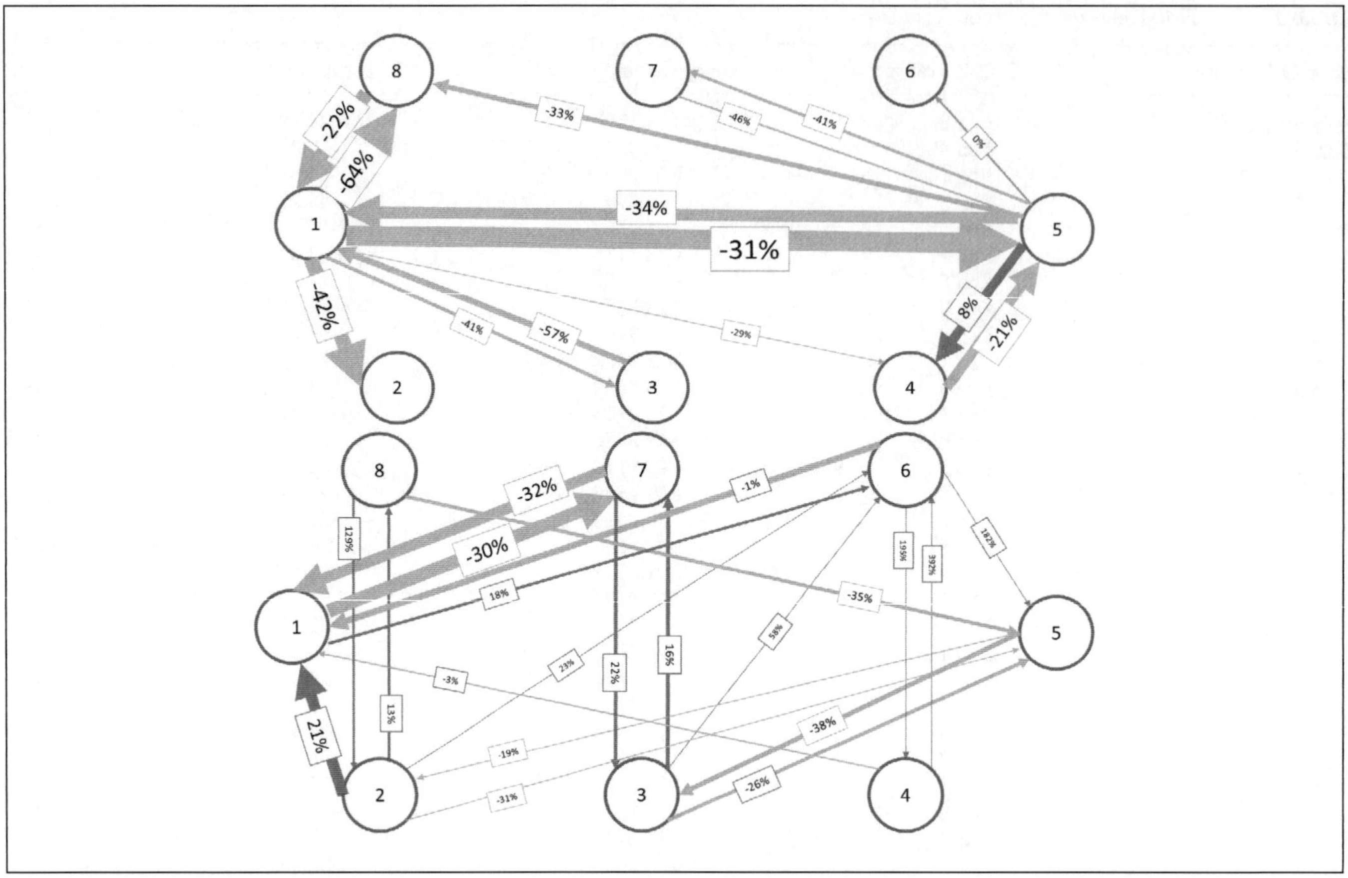

Figure 6. Percent change in travel time with consolidated intersection design.

Table 3. Overall Effectiveness of Consolidated Intersection Design

Overall effectiveness based on travel time reduction		Commercial development speed	
	In-network	Assume 15 mph	Assume 10 mph
Weighted average change in travel time	−31.9 s/veh	−16.2 s/veh	−8.3 s/veh
Percent change	−48.5%	−24.6%	−12.6%
Weighted average change in delay	−29.1 s/veh	NA	NA
Percent change	−76.4%	NA	NA

to drive in each of the two models if the origin and destination of the driver are the same. The O-D routes were derived using the same labels as shown in Figure 2.

A statistical analysis was performed separately for travel time and delay to test for levels of significant change. A two-way ANOVA was fit to the data in which condition (existing or proposed) and route were factors. Two-sided pairwise comparisons (using Bonferroni's adjustment for multiple comparisons) were made to compare base design routes with related proposed design routes. Pairwise comparisons between common O-D routes follow in Table 1 for travel time and Table 2 for delay. The results shown in these tables are discussed in the Results section.

Results

Some routes in the consolidated intersection are not displaced due to the proposed design, but some are. It is important to differentiate the two while evaluating the results. Twenty-four of the 44 O-D routes that contain traffic remain unchanged due to the geometric changes of the consolidated intersection design. Because these routes were not displaced, the change in travel time and delay can be evaluated directly from Tables 1 and 2 without needing to consider the effect of increased or decreased trip length.

With the exception of unsignalized right turns at the periphery of the study area, all non-displaced routes showed a statistically significant reduction in travel time and delay. Travel time and delay changes are reported for displaced routes, but these measures only take into account the "in-network" portion of the trips. In-network refers to travel within the boundaries of the simulation. The in-network reduction of travel time was 31.9 s/veh (–48.5%), and the in-network reduction of delay was 29.1 s/veh (–76.4%).

Displaced routes must travel within the commercial development, but the simulation did not model this. The travel time and delay of vehicles within commercial developments are highly variable, but some common ground must be established in order to compare the before condition to the proposed design. An average speed of 15 mph within the commercial development was used to adjust the travel time of displaced routes. Figure 6 shows the percent change in travel time for each O-D route in the study area, taking into account the 15 mph average speed within commercial developments. The thickness of the lines and the size of the label correlate with the volume of the O-D routes.

Assuming a 10 mph average speed within the commercial development is a more conservative assumption. Table 3 shows the overall changes in travel time and delay in the network, and also assuming average speeds within commercial developments of 10 and 15 mph.

Table 3 shows that even a conservative assumption of 10 mph results in an overall reduction in travel time of 8.3 s/veh (–12.6%).

Discussion

The scope of this research was on operational performance, so safety can only be discussed here in qualitative terms. The consolidated intersection design reduced conflict points from 96 in total amongst the three intersections in the base condition to only 18 in the proposed design. If a conflict point represents a point were conflicting vehicles could potentially collide, then it can be inferred that there is much less potential for collisions with the consolidated intersection design. Of course, many factors play into the occurrence of crashes beyond just the number of conflict points, but the 81% decrease in conflict points shows promise that the proposed design is not less safe than the conventional base design.

Bike lanes, sidewalks, crosswalks, and median refuge areas are facilities that may be considered with the consolidated intersection design. Bike lanes can be added parallel to the travel lanes. Sidewalks may be installed parallel to the roadways as well. There are two primary options when providing adequate crossings for pedestrians. Figure 7*a* shows a plan view of the Western intersection if pedestrian crossings were provided. Note that this same concept can be applied to the Eastern intersection.

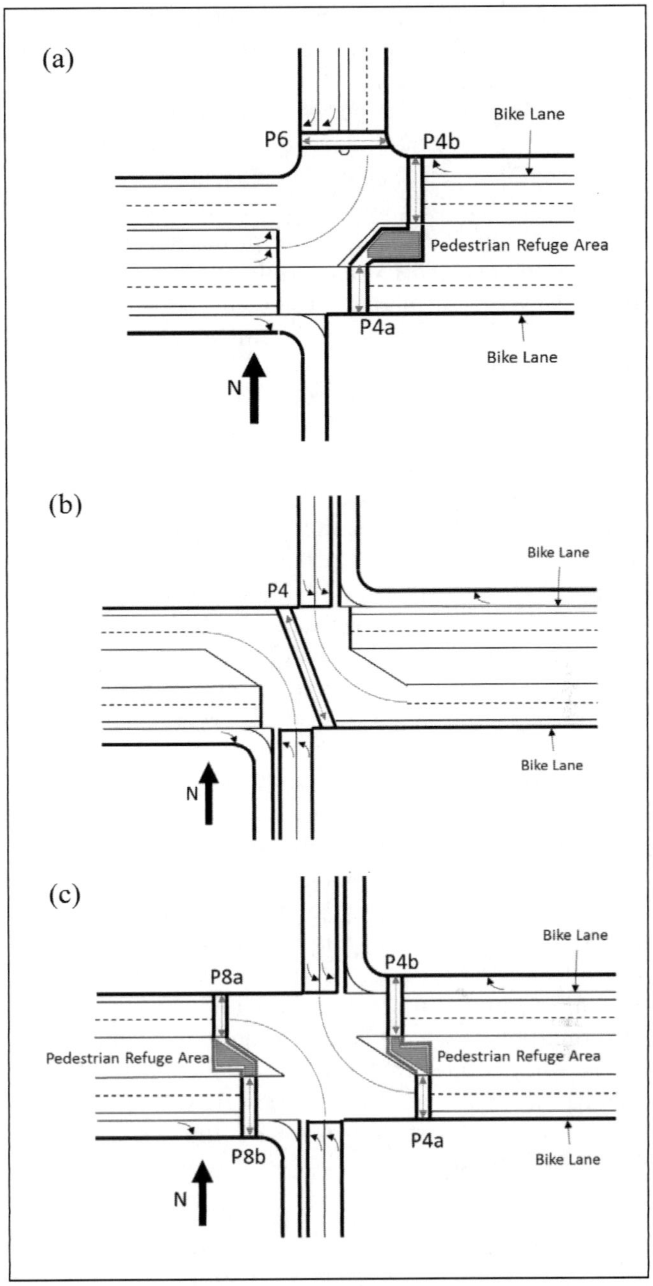

Figure 7. Pedestrian and bicyclist accommodations at (*a*) outer intersection and (*b*, *c*) center intersection.

As can be seen from Figure 7a, signal control must be installed to stop eastbound through vehicles to allow pedestrian phase P4a to operate. Otherwise, all other signal operations are the same as outlined previously in the report. P4a can run concurrently with the eastbound left/southbound right phase pair. P4b can also run concurrently with the same phase pair as P4a, but the eastbound through phase may remain green. A pedestrian refuge area between these two crosswalks should be built in the median to accommodate pedestrian storage. P6 can run concurrently with the eastbound through/westbound through phase pair. Also, an additional crosswalk may be added parallel to the eastbound lanes, crossing the south leg of the intersection.

Figure 7*b* presents a pedestrian crossing option for the Center intersection. This crosswalk traverses the entire arterial in one stage. The side-street approaches must be offset a little to provide enough space for the crosswalk to slip in between the opposing side-street left-turn movements. This option only works if sufficient right-of-way is available to offset the side-street approaches. Pedestrian phase P4 can only operate concurrently with the side-street left-turn phase pair. Signalized crosswalks may be added parallel to the arterial.

Figure 7*c* displays a second option for providing a signalized pedestrian crossing across the arterial. This option allows for two separate two-stage pedestrian crossings, each having a pedestrian refuge area in the median. Pedestrian phases P4b and P8b would operate concurrently with the side-street left-turn phase pair. The challenge with operating P4a and P8a is that these pedestrian phases conflict with every vehicular phase pair. These two pedestrian phases could operate only as an exclusive pedestrian phase at the intersection. If this were to occur, then P4b and P8b could operate during this exclusive pedestrian phase as well. Signalized crosswalks may be added parallel to the arterial.

Each commercial development is unique, and the required emergency access points must be determined prior to designing a consolidated intersection for a particular location. If emergency access is required at all three intersections in all directions (as provided in the base condition), then the design of the consolidated intersection must strategically use pavement markings and other delineators to channelize traffic flow, but still allow emergency vehicles to use pavement that was retained from the base design.

Conclusions

Two microsimulation models were compared to determine the operational effectiveness of the consolidated intersection design at a real-world site. The first model simulated base conditions of the study site, with traffic data collected during peak shopping demand. The second model simulated the consolidated intersection design for the same location. Travel time and delay were measured with the simulations, and a comparison test was conducted for each unique O-D route in the study area.

The overall reduction of in-network travel time is 31.9 s/veh (48.5%), and 29.1 s/veh (76.4%) for in-network delay. The consolidated intersection design heavily benefited travel for vehicles using the arterial as a through route. The arterial eastbound through route saw a travel-time reduction of 23 s, which was a decrease of 32%. The delay per vehicle in this direction fell by 23 s, which was a 75% reduction. The westbound through route had a travel-time reduction of 27 s, resulting in a decrease of 34%. The delay per vehicle in this direction dropped by 27 s, amounting to a 72% reduction.

All non-displaced routes, not including uncontrolled right-turn movements at the periphery of the study area, saw statistically significant reductions in travel time and delay. These results are a good sign that the consolidated intersection design maintains and improves all routes that do not have to travel additional distances due to the proposed design.

It is difficult to make clear-cut conclusions about how the consolidated intersection design affected the operations of the displaced routes. However, by assuming an average speed of travel within the commercial developments and assuming that displaced vehicles travel the full distance required to meet the original O-D route, an adjusted travel time can be calculated and compared with the travel times of the base design. The overall reduction in adjusted travel time, assuming a 15-mph speed of vehicles within commercial developments, is 16.2 s/veh, resulting in a 24.6% decrease in travel time. Even if a 10-mph speed is assumed, travel time is reduced by 8.3 s/veh (12.6%).

Based on the results of this research, the consolidated intersection design improves travel time and delay for arterials that run adjacent to commercial developments. Trips that either originate or conclude in the commercial developments are also positively affected by the consolidated intersection. Trips that both originate and conclude in the commercial developments may be displaced by a considerable distance, but more research is needed to clearly understand how this considerable displaced distance affects the travel time and delay of the routes within the commercial development.

Future Work

The 39th Street corridor was chosen as a real-world test case to verify that the consolidated intersection design could improve traffic operations for one particular location. Given the success of this trial application, the next step is to test the applicability of the consolidated intersection for various ranges of traffic volumes and varying demand on particular O-D routes within the study area.

Additional research should be conducted to investigate how displaced routes are affected in more detail to include travel time and delay within commercial developments.

Additional research should be conducted to understand the operational effects of deploying this proposed design within a larger signalized arterial network.

Because the signals in the consolidated intersection design are two-critical-phased, it is practical to consider half-cycling these intersections in relation to adjacent signalized intersections on the arterial.

Author Contributions

The author confirms contribution to the paper as follows: study conception and design: DC; data collection: DC; analysis and interpretation of results: DC; draft manuscript preparation: DC.

References

1. *Highway Capacity Manual: A Guide for Multimodal Mobility Analysis*, 6th ed. Transportation Research Board, Washington, D.C., 2016.
2. Hummer, J., B. Ray, A. Daleiden, P. Jenior, and J. Knudsen. *Restricted Crossing U-Turn Informational Guide*. Report No. FHWA-SA-14-070. Federal Highway Administration, Washington, D.C., 2014.
3. Hughes, W., R. Jagannathan, D. Sengupta, J. Hummer, and M. Smith. *Alternative Intersections/Interchanges: Informational Report (AIIR)*. Report No. FHWA-HRT-09-060. Federal Highway Administration, Washington, D.C., 2009.
4. Steyn, H., Z. Bugg, B. Ray, A. Daleiden, P. Jenior, and J. Knudsen. *Displaced Left Turn Intersection Informational Guide*. Report No. FHWA-SA-14-068. Federal Highway Administration, Washington, D.C., 2014.
5. Park, B., and J. Won. *Microscopic Simulation Model Calibration and Validation Handbook*. Report No. FHWA/VTRC 07-CR6. Virginia Transportation Research Council, Charlottesville, VA, 2006.
6. *Protocol for Vissim Simulation*. Washington State Department of Transportation, Olympia, WA, 2014.

The Standing Committee on Operational Effects of Geometrics (AHB65) peer-reviewed this paper (18-04713).

Article

TRR
JOURNAL OF THE TRANSPORTATION RESEARCH BOARD

Transportation Research Record
2018, Vol. 2672(17) 108–119

DOI: 10.1177/0361198118792755
journals.sagepub.com/home/trr
SAGE

Operational Effects of the Displaced Partial Cloverleaf Interchange

Larry F. Sutherland[1], Daniel J. Cook[2], and Karen K. Dixon[3]

Abstract

As the roadway infrastructure in the United States evolves, transportation agencies continue to seek effective interchange alternatives that can accommodate site-specific needs, such as high through and turning volumes, or optimized traffic signal operations (by minimizing the number of traffic signal phases). In recent years, innovative interchanges such as the diverging diamond have enabled transportation agencies to explore unique designs compatible with their specific needs. This paper introduces a new innovative interchange, known as a displaced partial cloverleaf (DPC) interchange. This unique interchange is characterized by a single intersection, six free-flow movements, only four movements controlled by the solitary traffic signal, an intersection location that can be shifted, and only 12 conflict points. The flexible intersection location can help an agency improve corridor progression, eliminate conflicting queues from nearby intersections, or enhance/maintain the bridge structure without compromising the intersection operations. A microsimulation study was conducted to compare the operational performance of a DPC interchange to that of a four-quadrant type B partial cloverleaf (PARCLO B-4Q) interchange, which is very similar in layout to a DPC interchange. Results show significant decreases in experienced travel time for left-turning vehicles with the DPC. The DPC interchange is recommended as an alternative option for interchanges with very heavy left-turning onramp demand in combination with heavy crossroad demand.

Transportation agencies have been looking for effective interchange alternatives to improve the performance of the roadway infrastructure. Innovative interchanges such as the diverging diamond interchange improve the efficiency of the interchange while still using the existing footprint of the interchange. This paper introduces a new innovative interchange, known as a displaced partial cloverleaf (DPC) interchange. The DPC combines benefits common to several traditional and innovative interchanges resulting in a unique interchange design. This paper briefly reviews advantages and disadvantages of known interchange configurations and compares the operational performance of the DPC to that of a PARCLO B-4Q using a microsimulation study.

Literature Review

Every interchange design is characterized by unique features that customize the interchange to the location of interest. In some cases, limited right-of-way widths can restrict the width of the interchange footprint. In other cases, the design of loop ramps can vary substantially depending on the available right-of-way, the expected operating speed, or the presence of collector–distributor roads or heavy vehicles. Though a wide variety of interchanges can be considered, this paper reviews interchange types that are likely to be considered for locations where a DPC may also be an option. These alternative interchanges include, but are not limited to, the following:

- Diamond;
- Partial cloverleaf (PARCLO) – type A and B;
- Single-point urban interchange (SPUI); and
- Diverging diamond.

Figure 1. presents a variety of diamond and partial cloverleaf interchanges.

Diamond

Historically, transportation agencies have constructed diamond interchanges (*2*) at urban service interchange

[1]WSP, Columbus, OH
[2]MRIGlobal, Kansas City, MO
[3]Texas A&M Transportation Institute, College Station, TX

Corresponding Author:
Address correspondence to Daniel J. Cook: dcook@mriglobal.org

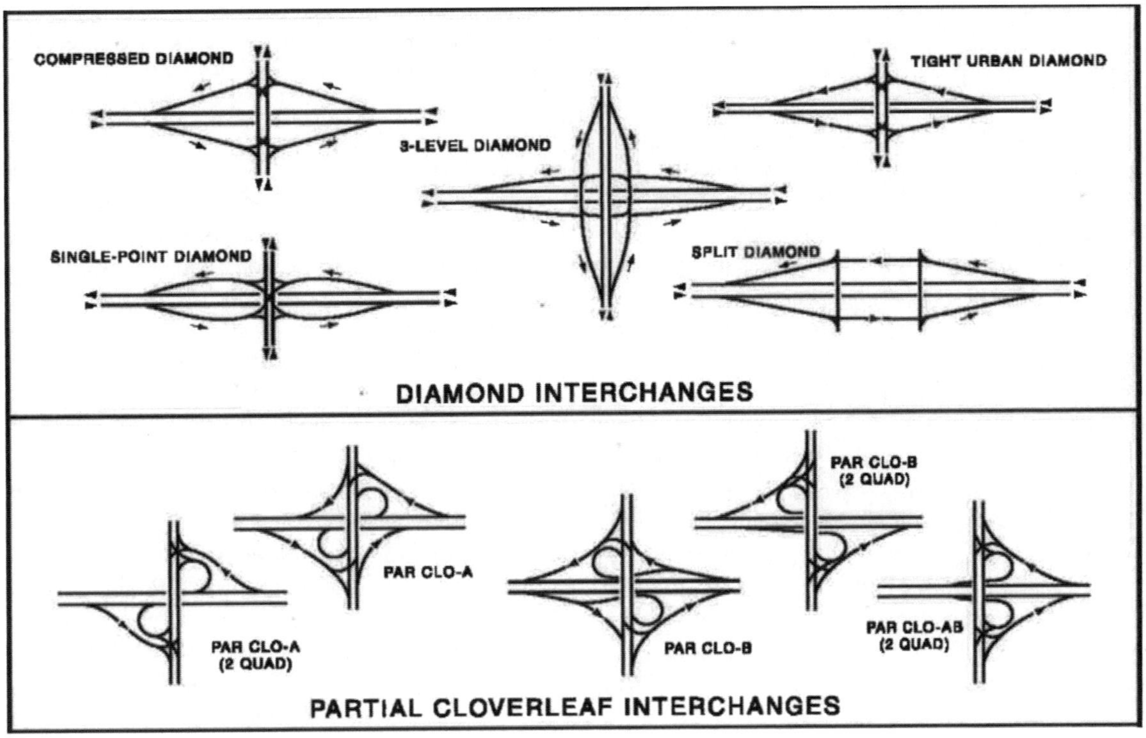

Figure 1. Varieties of diamond and partial cloverleaf interchanges (*1*).

locations. The diamond may have stop-controlled terminal intersections for low-volume locations, but most often includes two signalized terminal intersections operating with three critical phases each on the crossroad when warranted. In many locations, however, the diamond interchange is located on a roadway with increasingly large traffic volumes. A crossroad driver may experience delays if traveling along the corridor during congested conditions. This is often caused by limited queue storage on the bridge or high left-turn volumes that spillover to upstream interchanges (*2*). A tight urban diamond interchange is a specific type of diamond interchange used in conjunction with freeway frontage roads, in which the two intersections are 250–350 ft apart.

Partial Cloverleaf

The PARCLO (*2*) is similar to the diamond but characterized by the addition of loop ramps. This design can accommodate unique land constraints (railroads parallel to the freeway or crossroad, wetlands, and so forth) and does not require drivers to stop for many of the movements. The PARCLO may include a type A (where the loop ramps are located in **A**dvance of the bridge on the crossroad) or a type B (loop ramps are located **B**ehind the bridge on the crossroad). The footprint of the PARCLO is generally larger than that of the diamond. This extra right-of-way is needed to provide adequate space for loop ramps.

Partial Cloverleaf Type B-4Q (PARCLO B-4Q)

The PARCLO B-4Q interchange is less utilized in the United States versus other PARCLO types. However, the PARCLO B-4Q is ideal for use with crossroads where bi-directional progression is the goal. Each signalized ramp terminal only controls one direction of travel on the crossroad. This is ideal for achieving signal progression. Even on superstreets where each direction of travel may be controlled by independent cycle lengths, the PARCLO B-4Q can accommodate independent cycle lengths for each direction of travel.

Single-Point Urban Interchange

The SPUI (*3*) has a footprint size similar to a tight urban diamond, but instead of having two signalized terminal intersections, the crossroad turning maneuvers occur at a single signalized location. The single-point intersection is positioned directly above or below the highway, and this configuration generally requires a significantly larger structure than is typically used for a diamond or a

PARCLO. The single signalized intersection consists of three critical phases.

Diverging Diamond

The diverging diamond (*4*) is an innovative interchange that can be used at locations with high left-turn volumes onto the highway. This design requires the traffic to cross to the opposing (left) side of the road at the first crossroad terminal intersection. The left-turn traffic can then execute a free-flow left-turn onto the entrance ramp at the next crossroad terminal intersection, or return to the right side of the road. Both terminal intersection traffic signals operate with two critical phases, resulting in both reduced lost time and delay than can be expected at traffic signals with multiple phases. A diamond interchange can be retrofitted into a diverging diamond by utilizing the existing structure and reconstructing the approaches at the terminal intersections (*5*). The terminal intersections are located adjacent to the highway off-ramp locations. The design does introduce a challenge for pedestrians who must either cross to the median area or find themselves in direct conflict with the free-flow left-turning vehicles.

Operational Evaluation of Interchanges

The *Highway Capacity Manual* (HCM) (*6*) provides a multimodal approach to quantifying the operational performance of ramp terminals and alternative intersections in chapter 23. For each movement, experienced travel time (ETT) is calculated as the sum of control delay and extra distance travel time incurred by each movement. ETT is used to determine the level of service for each movement. There is no one overall level of service score or parameter assigned to the interchange as a whole.

Design of the DPC Interchange

The DPC incorporates a single two-critical-phase traffic signal, and other characteristics common to the PARCLO (loop and free-flow configuration). The DPC is similar to a SPUI in that it uses a single intersection, and a diverging diamond in that it prioritizes left-turn maneuvers. As depicted in Figure 2, the freeway runs up and down the page, and the crossroad runs from left to right. The signalized intersection is shown in the immediate vicinity of the structure; however, the DPC configuration enables the single signalized intersection to shift along the crossroad. The shifting of the signalized intersection can better enable progression, or can eliminate conflicting queues between this intersection and a closely spaced adjacent intersection. In some cases, the signalized intersection could shift by as much as 1,000 ft. The interchange footprint is similar to that of a PARCLO B-4Q, but the DPC crosses over the signalized left-turn upstream of the opposing left-turn movement, thus eliminating the second signal for each left-turn movement.

Inspection of Figure 2 demonstrates how the combined use of loop ramps, collector roads, left-turn priority design, and a two-critical-phase traffic signal accompanied by six free-flow movements enables the DPC to perform well at locations with high turn volumes or sites where the single terminal intersection cannot be conveniently located under an existing structure. Table 1 contrasts several features of the DPC to other interchange base configurations. It is notable that the number of free-flow movements is greater for the DPC, which also maintains a two-critical-phase traffic signal. In addition, the DPC only has 12 conflict points, the smallest number for any known service interchange. Crashes generally occur where conflict points exist, so typically fewer conflict points implies fewer crashes.

As previously mentioned, the DPC layout is very similar to a PARCLO B-4Q. In fact, the DPC could be used to replace a PARCLO B-4Q while still maintaining the advantages that a PARCLO B-4Q has for providing great bi-directional progression within a coordinated signal system. A wider cross-section may need to be provided, depending on the location of the intersection and number of left-turn lanes required at the intersection.

Riniker proposed a similar innovative interchange design called the *folded interchange* that uses contraflow left-turn lanes within the median of the crossroad (*7*). The folded interchange was analyzed as a replacement for a full cloverleaf interchange. The contraflow left-turns essentially allowed the left-turning movements to bypass the first terminal signal. The DPC interchange performs a similar function by crossing the left-turn movement across the opposing crossroad direction before the crossing of the opposing left-turn movement.

Microsimulation Study

A microsimulation study was conducted using Vissim 8 to compare the operational effects of a DPC interchange to a PARCLO B-4Q interchange. A real-world PARCLO B-4Q interchange was used as the basis for the geometrics of the microsimulation model. This PARCLO B-4Q interchange was modified into a DPC interchange as a separate microsimulation model.

PARCLO B-4Q Model

The PARCLO B-4Q microsimulation model is shown in Figure 3. This interchange is based on the I-526 and US-78 interchange in North Charleston, South Carolina. The mainline freeway was not modeled, because this

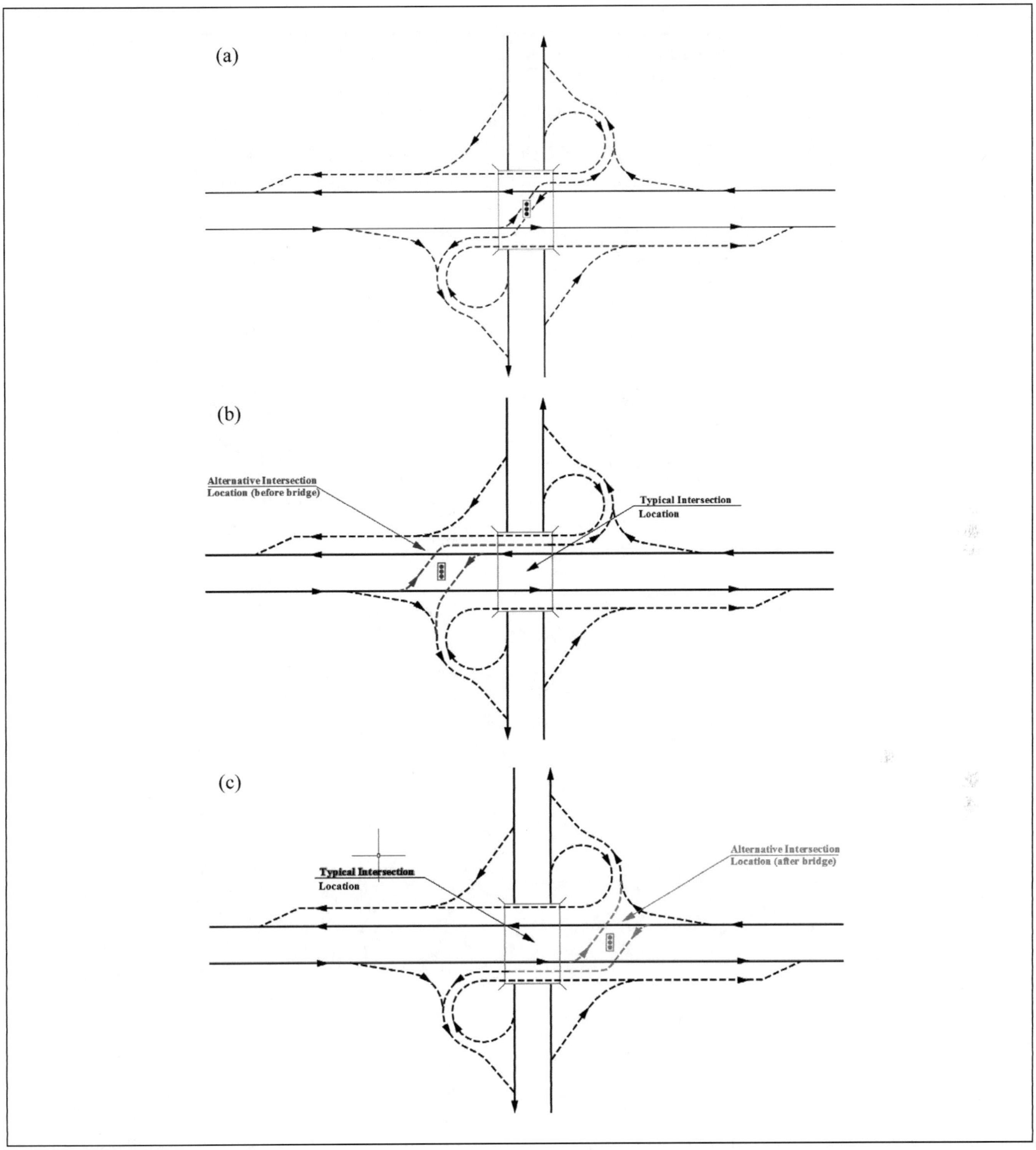

Figure 2. DPC schematic.

study is concerned with the operational performance of the crossroad and the interchange ramps. While not shown in Figure 3, the crossroad approaches extend 1,160 ft upstream of the first deceleration lane encountered. This excessive distance was chosen to allow the vehicles in the simulation enough distance and time to position themselves in the proper lane before the interchange (and before the point at which the travel time collection begins) depending on their destination. No field visits were made to the real-world interchange, so the

Table 1. Base Interchange Alternatives Comparison

Interchange type	Number of intersections	Free-flow movements[a]	Movements controlled by traffic signal	Intersection locations[c]	Number of conflict points	Number of critical signal phases
DPC	1	6	4	Variable	12	2
Diamond	2	0	10	Fixed	26	3
PARCLO A/B[b]	2/2	4/4	6/6	Fixed/fixed	12/12	2/2
SPUI	1	2	8	Fixed	18	3
Diverging Diamond	2	4	6	Fixed	14	2

[a]Designs can be modified to include additional free-flow movements.
[b]Loop ramps require more space than diamond ramps.
[c]Fixed intersection locations means intersections are fixed to one side of the freeway.

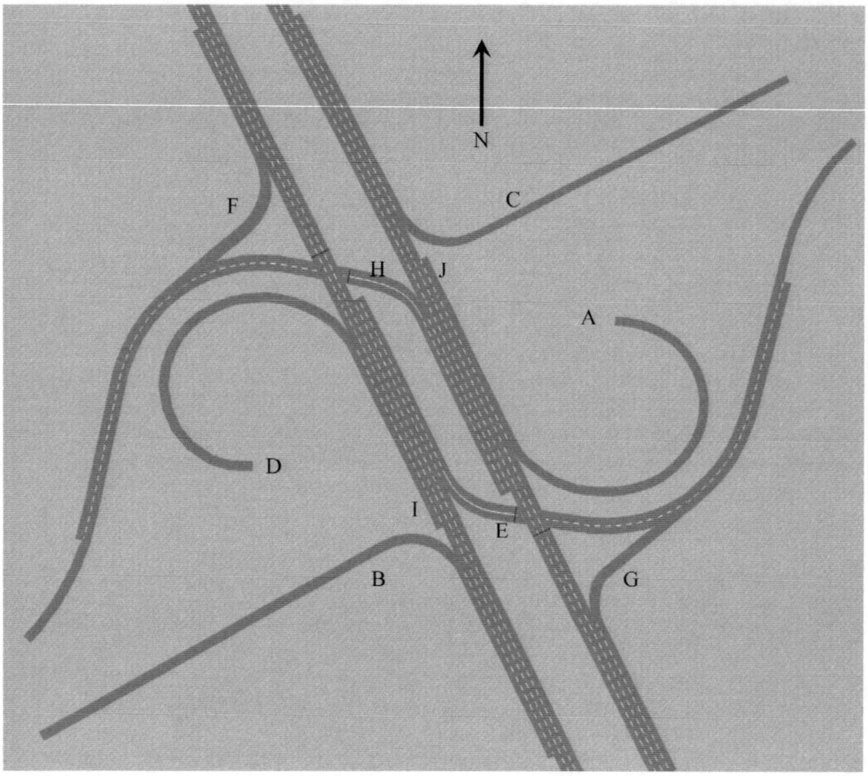

Figure 3. Microsimulation model of PARCLO B-4Q interchange.

model was not calibrated to match field conditions. However, many educated assumptions were made to calibrate the model to establish practical behavior. Table 2 presents these calibration assumptions in greater detail.

To ensure that the model was generating reasonable travel time averages, the simulation output was compared to calculated ETT values using HCM procedures. Because the O–D routes E and H are affected the most by the change from PARCLO B-4Q to DPC configuration, routes E and H were scrutinized the most during these validation steps. Using HCM procedures in chapter 23 of the *Highway Capacity Manual*, the ETT for O–D routes E and H of a PARCLO B-4Q are calculated to be 66.7 s and 66.4 s respectively using the demand inputs for scenario 1 (discussed in Table 4) and the travel time collection begin and end points (discussed in Table 3). These travel time collection begin and end points operate effectively as the cordon line drawn in Exhibit 23-5 in the HCM. The PARCLO B-4Q simulation model produces ETT averages of 66.0 s and 65.9 s for O–D routes E and

Table 2. Assumptions Used to Calibrate the PARCLO B-4Q Model

Calibration parameter	Value used and justification
Crossroad desired speed	45 mph (posted speed limit)
Loop ramp reduced speed	25 mph (ramp advisory speed)
Reduced speed of direct offramp curve	20 mph (ramp advisory speed)
Reduced speed of crossroad right-turns	20 mph (assumed based on curve radius)
Reduced speed of left-turn approach curve	20 mph (assumed based on curve radius)
Reduced speed of outer connection curve	35 mph (assumed based on curve radius)
Reduced speed of direct offramp curve	20 mph (assumed based on curve radius)
Percentage of heavy vehicles	2 percent

Table 3. Travel Time Collection Locations

O–D route	Begin travel time collection	End travel time collection
A	15 ft after the beginning of loop ramp	50 ft after the westbound direct offramp acceleration lane ends
B	15 ft after the beginning of direct offramp	40 ft after the eastbound direct offramp acceleration lane ends
C	15 ft after the beginning of direct offramp	50 ft after the westbound direct offramp acceleration lane ends
D	15 ft after the beginning of loop ramp	40 ft after the eastbound direct offramp acceleration lane ends
E	35 ft before the beginning of the southbound right-turn deceleration lane	10 ft before the end of the eastbound onramp
F	35 ft before the beginning of the southbound right-turn deceleration lane	35 ft before the end of the westbound onramp
G	55 ft before the beginning of the northbound right-turn deceleration lane	10 ft before the end of the eastbound onramp
H	55 ft before the beginning of the northbound right-turn deceleration lane	35 ft before the end of the westbound onramp
I	35 ft before the beginning of the southbound right-turn deceleration lane	40 ft after the eastbound direct offramp acceleration lane ends
J	55 ft before the beginning of the northbound right-turn deceleration lane	50 ft after the westbound direct offramp acceleration lane ends

H respectively. The similarity of the travel time values shows that the microsimulation model produces reasonable travel time estimates.

DPC Model

The DPC microsimulation model is shown in Figure 4. This model is a reconfiguration of the PARCLO B-4Q model discussed in the previous section. All calibrations shown in Table 2 remain in the DPC model except for the reduced speeds on the left-turn approach curves. The S-curve that left-turning vehicles must navigate through the DPC intersection was assigned a reduced speed of 20 mph.

Running Simulation Scenarios

Seven different demand scenarios were used to test the performance of the two models. Table 4 presents the demand volumes for each scenario as well as the optimized signal timings used for each scenario. All volumes are shown in units of vehicles per hour (veh/h). Note that all other unsignalized O–D routes have a constant demand of 400 veh/h for all scenarios. Table 5 shows signal timing parameters that remained constant through all the demand scenarios. The yellow and red intervals were calculated using procedures found in the *Guidelines for Determining Traffic Signal Change and Clearance Intervals* (*8*). The signals in each model were optimized as two separate two-phase rings; each left-turn crossing constituted a signal controller in the microsimulation. This was done so that each crossover can run its own optimized cycle and splits independent of the operation of the other crossover intersection. It should be noted that many modern controllers have the capability to run separate ring and barriers simultaneously; however, the ring barrier controller in Vissim is limited to one ring barrier structure, thus requiring the use of two signal

Table 4. Volume and Signal Inputs for Demand Scenarios

Demand scenario	Turning volumes at signalized ramp terminals	Maximum green settings for signalized ramp terminals	
		PARCLO B	DPC
1	1,000; 500; 500; 1,000	C = 35, Thru Split = 22, LT Split = 13 / C = 35, Thru Split = 22, LT Split = 13	C = 30, Thru Split = 17, LT Split = 13 / C = 30, Thru Split = 17, LT Split = 13
2	1,000; 500; 500; 2,000	C = 150, Thru Split = 115, LT Split = 35 / C = 35, Thru Split = 22, LT Split = 13	C = 40, Thru Split = 26, LT Split = 14 / C = 30, Thru Split = 17, LT Split = 13
3	2,000; 500; 500; 2,000	C = 150, Thru Split = 115, LT Split = 35 / C = 150, Thru Split = 115, LT Split = 35	C = 40, Thru Split = 26, LT Split = 14 / C = 40, Thru Split = 26, LT Split = 14
4	2,500; 500; 500; 2,500	C = 150, Thru Split = 118, LT Split = 32 / C = 150, Thru Split = 118, LT Split = 32	C = 60, Thru Split = 43, LT Split = 17 / C = 60, Thru Split = 43, LT Split = 17
5	1,000; 750; 750; 1,000	C = 85, Thru Split = 52, LT Split = 33 / C = 85, Thru Split = 52, LT Split = 33	C = 30, Thru Split = 16, LT Split = 14 / C = 30, Thru Split = 16, LT Split = 14
6	1,000; 250; 1,000; 2,000	C = 150, Thru Split = 130, LT Split = 20 / C = 35, Thru Split = 17, LT Split = 18	C = 35, Thru Split = 24, LT Split = 11 / C = 35, Thru Split = 17, LT Split = 18
7	1,000; 1,000; 250; 2,000	C = 135, Thru Split = 82, LT Split = 53 / C = 45, Thru Split = 34, LT Split = 11	C = 85, Thru Split = 48, LT Split = 37 / C = 30, Thru Split = 18, LT Split = 12

controllers in Vissim. This operation is especially desired on arterials where bi-directional progression is sought. This freedom is even more beneficial on superstreets where each direction of travel may be progressed using independent cycle lengths. The target volume to capacity ratio for each lane group was 0.85 to optimize the signal cycle and phase durations. Each intersection ran its signal controller in a fully actuated free mode. Each phase was assigned a 2-s extension to green time. The geometric changes between the two interchange types mean that optimized cycle lengths and phase durations can be drastically different for the same demand levels.

The purpose of comparing the operational performance of a DPC interchange to a PARCLO B-4Q interchange at different demand levels was to get an idea of how well, or how poorly, the DPC interchange could work as a suitable replacement for the PARCLO B-4Q. Seven demand scenarios were developed to represent a

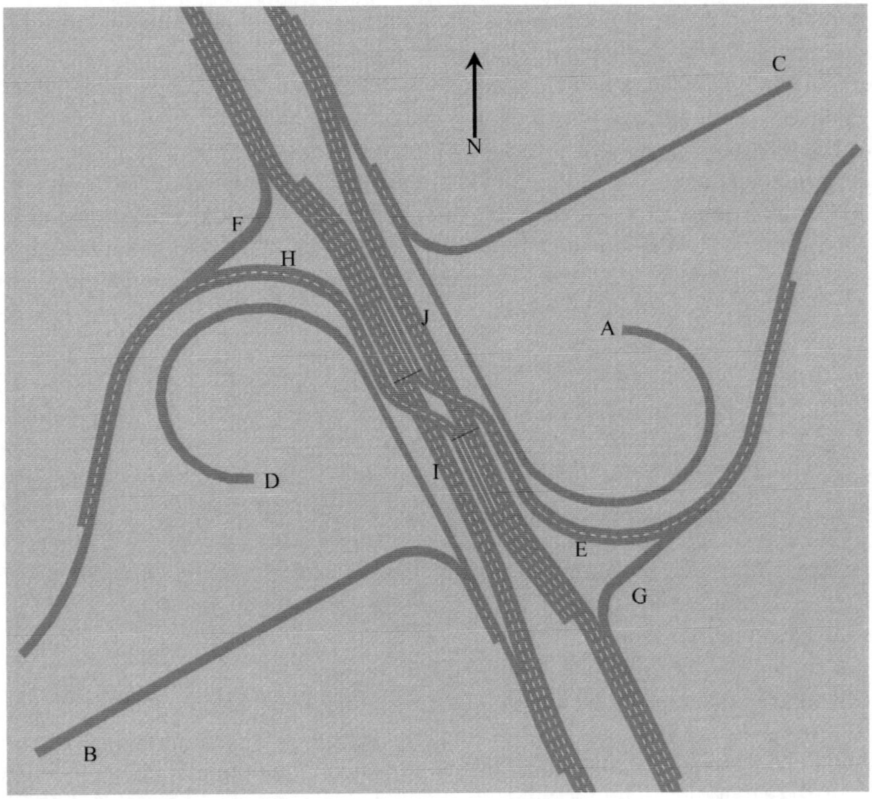

Figure 4. Microsimulation model of DPC interchange.

Table 5. Fixed Signal Inputs

	PARCLO B-4Q model		DPC model	
	Yellow duration (s)	Red duration (s)	Yellow duration (s)	Red duration (s)
Northbound through	5	1	5	1
Southbound left-turn	3	2	3	3
Southbound through	5	1	5	1
Northbound left-turn	3	2	3	3

variety of practical demand levels. The number of possible demand scenarios is practically limitless, but for the purposes of this proof of concept, the study focused on these seven scenarios. Additional demand scenarios can be tested in future analyses. Each scenario is discussed in more detail below.

- Scenario 1: The crossroad carries moderate but equal demand through the interchange for both directions of travel. The left-turn demand is moderate but equal for both directions of travel.
- Scenario 2: The crossroad carries twice as many vehicles in the northbound direction compared to the southbound direction. The left-turn demand is moderate but equal for both directions of travel.
- Scenario 3: The crossroad carries heavy but equal demand through the interchange for both directions of travel. The left-turn demand is moderate but equal for both directions of travel.
- Scenario 4: The crossroad carries very heavy but equal demand through the interchange for both directions of travel. The left-turn demand is moderate but equal for both directions of travel. The signalized intersection demand at each ramp terminal is nearing capacity.
- Scenario 5: The crossroad carries moderate but equal demand through the interchange for both directions of travel. The left-turn demand is moderate to heavy but equal for both directions of travel.

- Scenario 6: The crossroad carries twice as many vehicles in the northbound direction compared to the southbound direction. The northbound left-turn is very heavy. This heavy left-turn demand causes highly skewed lane utilization of the through lanes at the upstream signalized intersection. The southbound left-turn demand is low. Even with the low southbound left-turn demand, the unequal lane utilization of the through lanes at the southern signalized intersection decreases the saturation flow rate of the through lane phase enough to create near-capacity conditions.
- Scenario 7: The crossroad carries twice as many vehicles in the northbound direction compared to the southbound direction. The southbound left-turn is very heavy, while the northbound left-turn demand is low.

Each model was simulated 20 times per scenario using different random seeds. Each simulation ran for 4,500 s; however, travel time data were only collected during the final 3,600 s to allow the simulation to establish desired conditions. Travel time was collected for each vehicle in the simulation along 10 different O–D routes. The letters used to identify these O–D routes are consistent with those found in Exhibit 23-20 of the HCM (*6*). The location of the start and end travel time collection points were the same in each model and are presented in Table 3.

There were some minor changes in distance traveled between collection points for the O–D routes because of interchange reconfiguration. These changes in distance traveled, though, are reflected in the average travel time statistic. Analyzing ETT is consistent with how the HCM quantifies operational performance of interchanges.

Statistical Analysis of Results

The average travel times for each of the 10 O–D routes from the PARCLO B-4Q model were compared to the same 10 O–D routes from the DPC model for each demand scenario. A t-test for independent samples with unequal variance was used to test for a statistically significant difference between the average travel times at the 95% confidence level. The null hypothesis of the test was that the difference in average travel time of O–D route i between the two interchange types equals zero. The statistical test was conducted to prove the null hypothesis wrong with 95% certainty. Even though the travel time distributions per O–D route are not normally distributed, the Central Limit Theorem (*9*) states the sampling distribution of average travel times will be approximately normal when the number of observations is large.

The results of the statistical analysis are shown in Table 6. Because this was a two-sided t-test, a change in average travel time is statistically significant at the 95% confidence level when the p-value is less than or equal to 0.025.

As shown in Table 6, there were 70 separate hypothesis tests conducted. The null hypothesis that the average travel times for O–D route i are equal between the DPC and PARCLO B-4Q interchange was rejected for 64 of the tests at a 95% confidence level. In other words, 64 of the tests showed that the average travel times for O–D route i were significantly different across the seven demand scenarios between the two interchange configurations at a 95% confidence level.

Table 7 presents the change in average travel time for all 10 O–D routes combined. Seven additional hypothesis tests were conducted analogous to the tests conducted for individual O–D routes. For all seven demand scenarios, the average travel times of all DPC routes combined were significantly decreased at the 95% confidence level as compared to the average travel times of all PARCLO B-4Q interchange routes.

Left-Turns from the Crossroad

In each of the seven demand scenarios, the average travel time decreased significantly for left-turns from the crossroad when comparing the DPC interchange to the PARCLO B-4Q interchange. This result was expected, because the left-turning movements in the DPC interchange no longer have to travel through two signalized intersections. This results in less control delay. Also, since the left-turns are removed from one of the two signalized intersections, there are fewer total vehicles using each intersection, which lowers the critical flow ratio for the intersection. Lower critical flow ratios lead to lower cycle lengths and shorter splits, which in turn produce lower control delays.

For the PARCLO B-4Q, the presence of the left-turning demand upstream of the crossover signal produces highly skewed lane utilization in the three through lanes of the upstream signalized intersection. Most of the vehicles that will turn left at the downstream crossover are positioned in the left-most through lane at the upstream intersection. A highly skewed lane utilization will lead to lower saturation flow rates and longer phase durations. The DPC interchange eliminates this highly skewed lane utilization, because the left-turning demand no longer has to use the through lanes (there is no longer an upstream signalized intersection).

Lastly, another factor that seems to have affected the decrease in ETT was how vehicles merged at the point when the two-lane onramp merged to one lane. When the maximum green settings were larger in the PARCLO B-4Q, large platoons were released from the crossover signals and inundated this merge point with many

Table 6. Results of Average Travel Time Comparisons

Demand scenario		EB to NB	EB to SB	WB to NB	WB to SB	SBLT	SBRT	NBRT	NBLT	SB	NB
		A	B	C	D	E	F	G	H	I	J
1	ΔETT[a]	0.4	1.2	1.2	0.5	−14.0	0.2	−0.3	−12.9	0.4	1.1
	Percentage change	1.3%	4.0%	4.3%	1.5%	−21.2%	0.7%	−1.0%	−19.6%	1.4%	3.4%
	p-value	<0.0001	<0.0001	<0.0001	<0.0001	<0.0001	0.0235	0.0011	<0.0001	<0.0001	<0.0001
	95% CI	Yes	Yes	Yes	Yes	Yes	Yes	Yes	Yes	Yes	Yes
2	ΔETT[a]	1.64	1.22	2.85	0.51	−16.20	0.16	−1.35	−11.31	1.30	−0.46
	Percentage change	4.7%	4.1%	9.9%	1.5%	−21.9%	0.6%	−4.4%	−17.5%	4.0%	−1.3%
	p-value	<0.0001	<0.0001	<0.0001	<0.0001	<0.0001	0.0527	<0.0001	<0.0001	<0.0001	0.0031
	95% CI	Yes	Yes	Yes	Yes	Yes	No	Yes	Yes	Yes	Yes
3	ΔETT[a]	1.70	2.45	2.84	1.28	−16.26	−0.34	−1.29	−15.69	−0.18	−0.26
	Percentage change	4.8%	8.2%	9.9%	3.7%	−21.9%	−1.2%	−4.2%	−21.3%	−0.5%	−0.7%
	I-value	<0.0001	<0.0001	<0.0001	<0.0001	<0.0001	0.0017	<0.0001	<0.0001	0.077	0.0138
	95% CI	Yes	Yes	Yes	Yes	Yes	Yes	Yes	Yes	No	Yes
4	ΔETT[a]	3.81	4.04	5.33	2.45	−20.94	−0.84	−1.73	−19.46	−0.22	−0.49
	Percentage change	10.7%	13.4%	18.4%	7.0%	−24.5%	−2.7%	−5.4%	−23.3%	−0.6%	−1.3%
	p-value	<0.0001	<0.0001	<0.0001	<0.0001	<0.0001	<0.0001	<0.0001	<0.0001	0.0781	0.0016
	95% CI	Yes	Yes	Yes	Yes	Yes	Yes	Yes	Yes	No	Yes
5	ΔETT[a]	0.47	1.18	1.30	0.46	−17.18	−1.88	−2.74	−16.30	−0.36	0.11
	Percentage change	1.4%	4.0%	4.6%	1.4%	−24.3%	−5.8%	−8.1%	−23.2%	−1.0%	0.3%
	p-value	<0.0001	<0.0001	<0.0001	<0.0001	<0.0001	<0.0001	<0.0001	<0.0001	0.0039	0.2297
	95% CI	Yes	Yes	Yes	Yes	Yes	Yes	Yes	Yes	Yes	No
6	ΔETT[a]	1.63	1.32	2.87	0.57	−28.16	−1.27	−0.45	−9.65	1.20	−0.03
	Percentage change	4.6%	4.5%	10.0%	1.7%	−33.4%	−3.7%	−1.6%	−14.7%	3.3%	−0.1%
	p-value	<0.0001	<0.0001	<0.0001	<0.0001	<0.0001	0.0007	<0.0001	<0.0001	<0.0001	0.4097
	95% CI	Yes	Yes	Yes	Yes	Yes	Yes	Yes	Yes	Yes	No
7	ΔETT[a]	2.97	1.19	3.80	0.53	−13.96	0.38	−4.43	−20.68	1.36	0.55
	Percentage change	8.4%	4.0%	13.2%	1.6%	−17.1%	1.4%	−10.2%	−28.1%	4.5%	1.3%
	p-value	<0.0001	<0.0001	<0.0001	<0.0001	<0.0001	<0.0001	<0.0001	<0.0001	<0.0001	0.0286
	95% CI	Yes	Yes	Yes	Yes	Yes	Yes	Yes	Yes	Yes	No

Note: EB = eastbound; LT = left-turn; NB = northbound; RT = right-turn; SB = southbound; WB = westbound.
[a]ΔETT is the change in experienced travel time in seconds.

Table 7. Results of Average Travel Time Comparisons for All Routes Combined

All interchange routes	Demand scenario 1	2	3	4	5	6	7
ΔETT[a]	−2.0	−1.82	−1.92	−1.95	−4.48	−2.10	−2.31
Percentage change	−5.2%	−4.7%	−4.9%	−4.8%	−10.4%	−5.2%	−5.2%
p-value	<0.0001	<0.0001	<0.0001	<0.0001	<0.0001	<0.0001	<0.0001
95% CI	Yes	Yes	Yes	Yes	Yes	Yes	Yes

[a]ΔETT is the change in experienced travel time in seconds.

vehicles. However, with the shorter maximum green times in the DPC configuration, smaller left-turning platoons were released more frequently. This acted as a metering mechanism to allow the smaller-platoon vehicles to merge into one lane downstream of the signal quicker and more efficiently, thus lowering the ETT.

Through Movements on the Crossroad

In four of the seven, and five of the seven demand scenarios, the change in average travel time was statistically significant at the 95% confidence level for the northbound and southbound through routes, respectively. However, even for the statistically significant changes, the changes in average travel time are very small. The largest increase across all seven scenarios for both directions was a 1.36-s increase in travel time and the largest decrease across all seven scenarios for both directions was a 0.46-s decrease. These very small changes are negligible. The results, however, are expected. For both interchange configurations, arterial through movements only travel through one signalized intersection. The distance traveled for a through vehicle did increase slightly with the new interchange design (because of the two S-curves through the interchange); but considering all changes as a whole, the simulation models show the through movements neither benefit nor suffer from the DPC interchange configuration.

Unsignalized Movements

For the most part, any change in the average travel time of unsignalized movements has more to do with the change in geometrics than the change in signalization of the ramp terminals. For example, traffic from the two direct offramps has to merge twice in the DPC configuration and only once with the PARCLO B-4Q configuration.

The direct offramps of the DPC consisted of O–D routes B and C. For all seven scenarios, the average travel time increase was statistically significant at the 95% confidence level. As the crossroad through demand increased, specifically in scenarios 3 and 4, the average travel time increased more as well. Intuitively, this makes sense; vehicles merging from the offramps will have fewer gaps to merge as the crossroad demand increases. Overall, though, the greatest increase in average travel time was 5.33 s, not a huge change, but big enough to produce an 18.4% increase over the PARCLO B-4Q interchange.

The loop ramps of the DPC consisted of O–D routes A and D. For all seven scenarios, the average travel time increase was statistically significant at the 95% confidence level. Similar to the direct offramps, as the crossroad through demand increased, the average travel time increased more as well. Intuitively, this makes sense for the same reasons stated in the previous paragraph. The greatest increase in average travel time was 3.81 s—again, not a big change, but big enough to produce a 10.7% increase over the PARCLO B-4Q interchange.

The channelized right-turns of the DPC from the crossroad onto the outer connection onramps consisted of O–D routes F and G. Most of the scenarios produced a significant decrease in average travel time, the greatest being a 4.43-s decrease for the northbound right-turn in scenario 7. The reasons for these changes are complex. Many factors can contribute to these changes, such as smaller and more frequent platoons being released onto the outer connection ramp that the right-turns are merging onto, thus leading to more frequent gaps for merging.

Conclusions

The DPC interchange was proposed as an alternative interchange at locations with high volumes of through and turning vehicles. The results of the microsimulation study show that the DPC may be a suitable replacement for the PARCLO B-4Q interchange. The operational benefit of the DPC increases when compared to the PARCLO B-4Q as through and left-turn demand increases. The operational changes of the through movements and unsignalized movements are small to negligible. Future research may be conducted to quantitatively compare the operational effectiveness of the DPC to other interchange types, such as the diamond interchange, SPUI, or diverging diamond interchange. However, the PARCLO B-4Q is designed to handle heavy through and left-turn volumes. The DPC outperformed the PARCLO B-4Q by significantly reducing ETT for left-turn movements without sacrificing the through movement performance, demonstrating that the DPC is a viable interchange alternative.

Another benefit of the DPC is that it can be placed in a signal system where each direction on the crossroad is progressed using independent cycle lengths. This study simply studied the DPC as a standalone interchange. The lower cycle lengths of the DPC can be fit into a coordinated network of signals as a half or third cycle. Future research should quantify the benefits of the DPC in a coordinated signal system as compared to other interchange configurations. Other future research needs include: economic analysis of the DPC; development of final designs accommodating a WB-67; queueing analysis; accommodations for non-motorist users; and further analysis of the merge point between the left-turn and right-turn traffic streams on the freeway onramp.

Author Contributions

The authors confirm contribution to the paper as follows: study conception and design: LFS, DJC, KKD; data collection: LFS; analysis and interpretation of results: DJC; draft manuscript preparation: LFS, DJC, KKD. All authors reviewed the results and approved the final version of the manuscript.

References

1. Leisch, J. P. *Freeway and Interchange Geometric Design Handbook*. Institute of Transportation Engineers, Washington, D.C., 2005.
2. American Association of State Highway Transportation Officials. *A Policy on Geometric Design of Highways and Streets*. AASHTO, Washington, D.C., 2011.
3. Messer, C. J., J. A. Bonneson, S. D. Anderson, and W. F. McFarland. *NCHRP Report: Single Point Urban Interchange Design and Operations Analysis*. Transportation Research Board of the National Academies, Washington, D.C., 1991.
4. Hughes, W., R. Jagannathan, D. Sengupta, and J. Hummer. *Alternative Intersections/Interchanges: Informational Report (AIIR)*. Report No. FHWA-HRT-09-060. Federal Highway Administration, U.S. Department of Transportation, Washington, D.C., 2010.
5. Schroeder, B., C. Cunningham, B. Ry, A. Daleiden, P. Jenior, J. Knudsen, and Kittelson & Associated, Inc. *Diverging Diamond Interchange Informational Guide*. Report No. FHWA-SA-14-067. Federal Highway Administration, U.S. Department of Transportation, Washington, D.C., 2014.
6. *Highway Capacity Manual: A Guide for Multimodal Mobility Analysis*, 6th ed. Transportation Research Board, Washington, D.C., 2016.
7. Riniker, K. A. The Folded Interchange: An Unconventional Design for the Reconstruction of Cloverleaf Interchanges. *ITE 2009 Technical Conference and Exhibit,* Session 15, Phoenix, AZ, 2009.
8. Noble, D. E. *Guidelines for Determining Traffic Signal Change and Clearance Intervals: An ITE Proposed Recommended Practice.* Institute of Transportation Engineers, Washington, D.C., 2015.
9. Ott, R. L., and M. Longnecker. *An Introduction to Statistical Methods and Data Analysis*, 6th ed. Brooks/Cole, Cengage Learning, Belmont, Calif., 2010.

The Standing Committee on Operational Effects of Geometrics (AHB65) peer-reviewed this paper (18-04726).

Article

TRR
JOURNAL OF THE TRANSPORTATION RESEARCH BOARD

Transportation Research Record
2018, Vol. 2672(17) 120–128

DOI: 10.1177/0361198118773505
journals.sagepub.com/home/trr
SAGE

Evaluating Safety Impacts of Access Management Alternatives with the Surrogate Safety Assessment Model

Kyung Min Kim[1], Mitsuru Saito[1], Grant G. Schultz[1], and Dennis L. Eggett[2]

Abstract

In a traditional safety impact analysis, it is necessary to have crash data on existing roadway conditions and a few years must pass before accumulating additional crash data to evaluate the safety impact of an improvement. This is a time-consuming approach and there remains uncertainty in the crash data integrity. The surrogate safety assessment model (SSAM) was developed for resolving these issues. With SSAM, a conflict analysis is performed in a simulated environment. A planned improvement alternative is modeled and no physical installation of the alternative is needed. This study evaluated if SSAM can be used to assess the safety of a highway segment in terms of the number and type of conflicts and to compare the safety effects of multiple access management alternatives. An evaluation of the effect of converting a two-way left-turn lane (TWLTL) into a raised median on a section of an urban street was performed using SSAM working on VISSIM simulation's trajectory files. The analysis showed that a raised median would be much safer than a TWLTL median for the same level of traffic volume, with approximately 32 to 50 percent reduction in the number of crossing conflicts. The analysis showed that about 34,000 to 38,000 veh/day would be the demand level where the median conversion is recommended for the four-lane study section. The study concluded that the combination of a simulation software program with SSAM could be a viable surrogate analysis approach for evaluating and comparing the safety effects of multiple access management alternatives.

The *Access Management Manual, Second Edition* defines access management as "the coordinated planning, regulation, and design of access between roadways and land development. It encompasses a range of access management alternatives that promote the efficient and safe movement of people and goods by reducing conflicts on the roadway system and at its interface with other modes of travel. These access management alternatives include improvements to benefit transit, pedestrians, and bicyclist, as well as different treatments for urban, suburban, and rural settings" (*1*).

One of the challenges with access management-related studies has been the availability of necessary data for conducting such studies. The Utah Department of Transportation (UDOT) began a light detection and ranging (LiDAR) data inventory in 2012 that includes access locations along all segments of the UDOT roadway network. The LiDAR inventory was repeated in 2014 and 2015. Combined with UDOT's crash data, the availability of access point data allowed an in-depth safety-related study to find relationships among access density, access management alternatives, and crash occurrences (*2*).

Safety impacts of access management alternatives can be analyzed using before-and-after studies. However, in a traditional safety impact analysis it is necessary to install an access management alternative in the field and a few years must pass to accumulate reliable crash data. This is a time-consuming study and there remains uncertainty in the crash data as a result of the random nature of crash occurrences. To circumvent the use of crash data a method called a conflict analysis was developed, which is performed typically in the field. Although this method is free from the issues pertinent to crash data, it still requires time and cost because the safety improvement measure must be installed at the study site. The surrogate safety assessment model (SSAM) was developed in an attempt to ameliorate the issues related to safety impact evaluation described above. With this method, a planned access management alternative is simply added to the simulation model and no physical installation of the alternative is needed. Thus, it requires less time and less cost.

In this study, SSAM was tested to determine whether it could be used to evaluate safety impacts of access management alternatives, specifically a two-way left-turn lane (TWLTL) and a raised median. This paper presents a brief

[1]Department of Civil & Environmental Engineering, Brigham Young University, Provo, UT
[2]Department of Statistics, Brigham Young University, Provo, UT

Corresponding Author:
Address correspondence to Mitsuru Saito: msaito@byu.edu

summary of the literature review, study methodology, calibration results, application of SSAM to safety impact evaluation, and conclusions.

Literature Review

The literature review of the study focused on traffic safety analysis methods using surrogate safety analysis measures. One of them is the traffic conflicts technique (TCT) and the other is SSAM. See Saito et al. (*3*) for an extensive literature review performed for the study.

TCT

The General Motors Research Laboratories (GMRL) originally developed the TCT in 1967 (*4*). TCT was performed in the field by observing and measuring crash potential on existing traffic facilities. Crash potential is defined as a conflict event, which is the occurrence of evasive vehicular actions and characterized by braking and weaving maneuvers forced by an impending collision or a traffic violation. Traffic violations are recorded as conflicts regardless of the presence of other vehicles. Five basic conflict categories were defined by GMRL including left-turn, weave, cross-traffic, red-light violation, and rear-end conflicts. The TCT has some strengths and limitations compared with the safety analysis based on actual crash records on specific facilities. Although the TCT analysis is free from the problems associated with the liability issues of crash records and the time required to wait for accumulating adequate sample sizes of crashes compared with the safety analysis method based on actual crash records, it still requires an existing facility that has enough traffic volume so that an adequate number of conflicts can be observed.

SSAM

SSAM was proposed by a research team at SIEMENS and was sponsored by the Federal Highway Administration. Gettman and Head (*5*) studied the potential for deriving surrogate safety measures from commonly available microscopic traffic simulation models, and their efforts eventually led to the development of the SSAM. Their study was followed by another study to test if SSAM could be used for practical engineering applications (*6*). The authors studied the relationship between simulated conflicts and actual crashes that took place at the studied intersections. They found that there was a significant relationship between the simulated conflicts and crashes (*7*).

SSAM is a post-processor of vehicle trajectory files produced by microscopic traffic simulation models. It operates by processing data describing the trajectories of vehicles driving through a simulated traffic facility and identifying conflicts. The vehicle trajectory input data for SSAM can be generated by traffic simulation software programs including VISSIM, Aimsun, Paramics, and TEXAS in a trajectory file format, specifically designed for SSAM. SSAM calculates surrogate measures of safety corresponding to each vehicle-to-vehicle interaction and determines whether or not each interaction satisfies the criteria to be deemed an official conflict. A table of all identified conflicts and their corresponding surrogate safety measures is then presented to the user (*8*). Conflicts identified by SSAM are crossing, rear-end, or lane-change movements, and the conflict angle determines conflict types as shown in Figure 1. The type is classified as a rear-end conflict if ||ConflictAngle|| <30 degrees, a crossing conflict if ||ConflictAngle|| >85 degrees, or a lane-change conflict otherwise. See Saito et al. (*3*) for a summary of surrogate safety parameters available in SSAM.

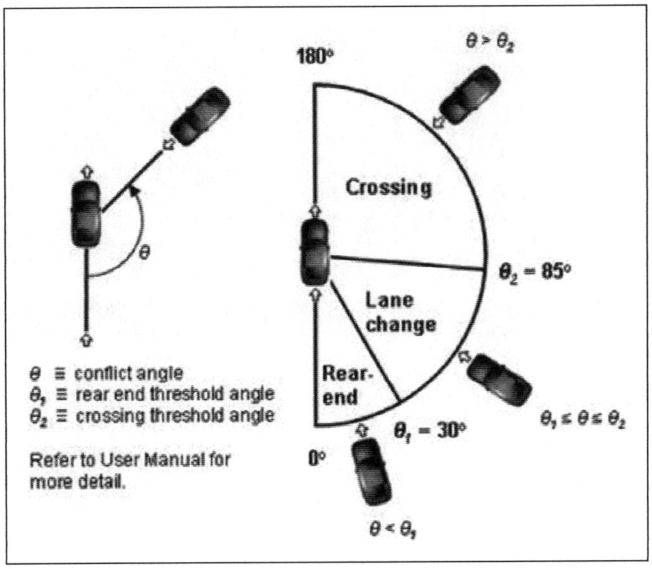

Figure 1. Conflict types by angle (*8*).

SSAM has been used in various traffic safety studies. For instance, Hummer et al. (*9*) evaluated operational, safety, and perceived effects of superstreets, called "restricted crossing U-turn intersections" using SSAM, and developed a level of service estimation program that could be used on North Carolina's urban and rural arterial roadway system. Al-Ghandour et al. (*10*) studied conflict patterns at single-lane roundabouts with and without slip lanes and compared their performances using VISSIM and SSAM. The SSAM analysis showed that conflicts in the merge area were more frequent than in the roundabout approach area, and that the installation of a free-flow slip-lane exit type reduced overall conflict occurrence. Lee et al. (*11*) investigated safety aspects of the cooperative vehicle infrastructure system (CVIS)-based urban traffic control system by applying SSAM. They reported that the number of rear-end conflict events decreased by 58 percent under the CVIS-based control. Huang et al. (*12*) conducted a study to identify if the VISSIM simulation model combined with SSAM could provide reasonable estimates of

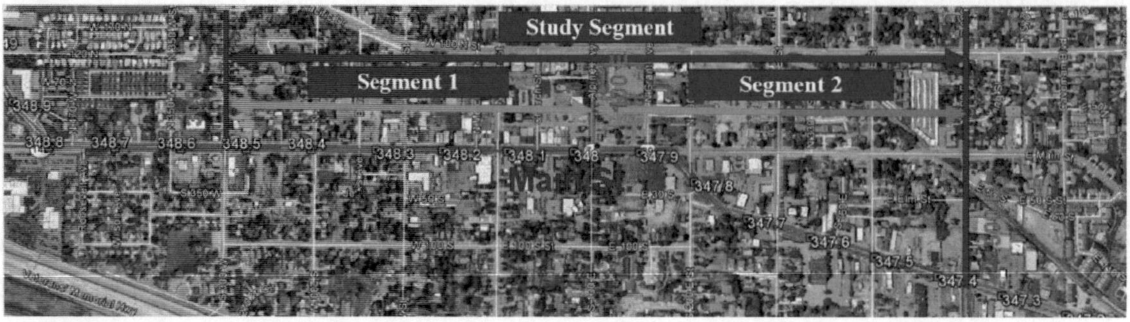

Figure 2. Main St. study section between 300 West and 500 East.

Table 1. Characteristics of Each Segment of the Study Section (*18*)

Segment[a,b]	Start point	End point	Length (mile)	Number of access	Access density (access/mile)	AADT
1	347.971	348.542	0.571	36	63.05	23,780
2	347.360	347.971	0.611	38	62.19	29,870
Section	347.360	348.542	1.182	74	62.61 (average)	26,825 (average)

[a]Speed limit on Segment 1: 35 mph between 300 West and 200 West and 30 mph between 200 West and 100 East; Segment 2: 35 mph from 100 East and 500 East.
[b] Lane width is 12.00 feet along the stretch. Shoulder width (that can be used for parking) is 8 ft for Segment 1 and 8–12 feet for Segment 2. Parking is prohibited throughout the stretch, except between 100 West to 100 East on Segment 1.

traffic conflicts that might take place at signalized intersections. They reported that results of data analysis showed that there was a reasonable goodness-of-fit between the simulated and the observed rear-end and total conflicts. However, they found that the simulated conflicts were not good indicators for the traffic conflicts generated by unexpected driving maneuvers such as illegal lane-changes in the real world. Dijkstra et al. (*13*) conducted a large-scale study, in which a 300-km^2 road network was modeled using Paramics. They collected six years of crash data from 569 pre-selected intersections in the road network. The study concluded that there was a significant statistical relationship between the observed crashes and the simulated conflicts.

Many more publications on SSAM application are available in the literature (*14–17*). Refer to Saito et al. (*3*) for a summary of these publications and others reviewed for the study. Although some studies have indicated that an overestimated number of conflicts were observed in the process, this literature review suggested that SSAM could be a viable tool to evaluate and compare safety effects of planned safety improvements on highways and intersections.

Study Methodology

The access management alternatives compared in this study included a TWLTL and a raised median. A section on Main St. (US-89) between 300 West and 500 East in American Fork was used because a 75-minute VISSIM model for this section was available from UDOT and the study section currently has a TWLTL. The study section is approximately 1.2 miles in length as shown in Figure 2. Both sides of the study section have various types of businesses and other land-use types. Because of the difference in average annual daily traffic (AADT), the study section was divided into two segments. Table 1 presents the characteristics of the two segments including the starting mile point, ending mile point, length, number of accesses, access density, and AADT. As shown in Table 1 the access densities of the two segments are similar, 62–63 accesses/mile, but their AADTs are quite different, which was the main reason to divide the study section into two segments.

The VISSIM model provided by UDOT did not contain all of the access driveways in the model. Therefore, all access driveways from the land uses along the study section were added to the model. Then, using the 8th Edition of the *Institute of Transportation Engineers (ITE) Trip Generation Manual* (*19*), potential trips generated by all the 74 access driveways for the 94 land uses located in the study section were added to the model.

In the first part of the task, before using SSAM for the safety evaluation, the Main St. VISSIM model was calibrated with 10 simulation runs by comparing the results of conflict frequencies and locations of the three types of conflicts with the actual crashes from the five-year crash data from year 2010 to 2014 (*18*). After the calibration work, safety impacts of a TWLTL and a raised median were evaluated. In the

Table 2. Trip Generation for Land Use on Main Street in American Fork (Samples)

Land use	Code name	ITE code	PM peak		
			Enter	Exit	Total
Emission Plus	Automobile Care Center	942	4	3	7
Office Complex	General Office	710	3	16	19
Timp Valley Floral	Specialty Retail Center	814	8	7	15
KFC	Fast Food with Drive-Through	934	69	63	132
Signs Now (Design – Signs)	Specialty Retail Center	814	16	13	29
Farmers Insurance Co.	General Office	710	1	4	5
Jalisco's Market (Supermarket)	Supermarket	850	32	28	60
Murdock & Searle (Dental Office)	Medical Dental	720	1	2	3
IMJ Therapy, INC.	General Office	710	1	2	3
NAPA Auto Parts	Auto Parts	843	19	19	38

second part of the task, a sensitivity analysis was conducted to identify a threshold AADT at which a TWLTL is recommended to be replaced with a raised median. For this analysis, the traffic volumes of Main St. and access roads (driveways) to the land uses along the segment were increased at 10 percent increments up to a 40 percent increase. A preliminary analysis showed that beyond the 40 percent increase in volume the VISSIM model started having vehicle entry problems. Thus, five levels of Main St. volume and five levels of access driveway volume were modeled for each access management alternative from 1.0 to 1.4 with an increment of 0.1 (10 percent). A total of 25 combinations of traffic volumes were simulated for each alternative, totaling 50 combinations for both alternatives. Because of this large number of simulation combinations, each case was simulated three times for a total of 150 simulation runs.

Data Preparation

The original model provided by UDOT needed modifications to include all the driveways along the study section because to test the safety effects of access management alternatives, demands from each of these driveways become important. Three of the four steps of the classical urban traffic demand forecasting method were followed.

In the trip generation step, traffic volumes (number of trips) were generated for the existing land uses along the study section using the 8th Edition of the *Trip Generation Manual* (*19*). Table 2 presents several sample calculations of the 95 land uses found along the study section with their land use Code names, ITE codes, and PM peak traffic volumes. Since the PM peak period was modeled in VISSIM, the average trip rate for weekday PM peak hour was estimated and then multiplied by the area of each land use to get total trips generated. Then, the generated trips were divided into entering and exiting trips based on directional distribution rate provided in the *Trip Generation Manual*. The first land use listed in Table 1, Emission Plus, is used to explain this process. The store has 1,661-ft^2 floor space. Its trip generation rate for the evening peak is 4.01 trips/1,000 ft^2. This results in 1,661 $ft^2 \times 4.01$ trip/hour/1,000 $ft^2 = 7$ veh/hour. The directional distribution is given as 51 percent entering, 49 percent exiting, resulting in 4 veh/hour entering and 3 veh/hour exiting, rounded to whole numbers to represent the number of vehicles as shown in the first data row of Table 2.

Because traffic volumes and vehicle compositions in traffic volume on Main St. are given in the files provided by UDOT, the mode split step was omitted. In the trip distribution step, all entering and exiting traffic volumes on each land use was distributed to the trip origin or destination in the traffic simulation model. The trips for each land use estimated from the trip generation step were considered entering and exiting from the through traffic volume on Main St. of the study section as the total traffic volume on Main St. cannot be changed.

After the trip distribution step, all entering and exiting trips were assigned to each access driveway. In the TWLTL case, all entering and exiting trips from each access driveway were assigned to each starting end of eastbound and westbound links on the Main St. study site. Some trips, which make a left turn to enter an access driveway or exiting from the access driveway in the TWLTL case, use the TWLTL on Main St. In the case of a raised median, the trips that made a left turn using the TWLTL in the TWLTL case would use the nearest signalized intersection downstream to make a U-turn to complete their trips to their destinations such as access driveway or the east or west ends of Main St.

Analysis Results

The access management measures compared in this study were a TWLTL and a raised median. This section presents the results of the calibration of SSAM for the study section, a comparison of the safety impacts of these two median measures, and the results of a sensitivity analysis to find threshold values for replacing a TWLTL with a raised median.

Table 3. Number of Crash Data Record and Conflicts Points

	No. of Crashes (2010–2014)				No. of Conflicts (75 min)			
Segment	Crossing	Rear end	Lane change	Total	Crossing	Rear end	Lane change	Total
1	30	49	0	79	84	764	158	1,006
2	73	67	8	148	192	1,880	434	2,506
Section summary	103	116	8	227	276	2,642	592	3,513

Table 4. Number of Crash Data Record and Relation Indices

	Relation Index = No. of Crashes/No. Conflicts			
Segment	Crossing	Rear end	Lane change	Total
1	0.355	0.064	0.000	0.078
2	0.380	0.036	0.018	0.059
Section summary	0.373	0.044	0.014	0.065

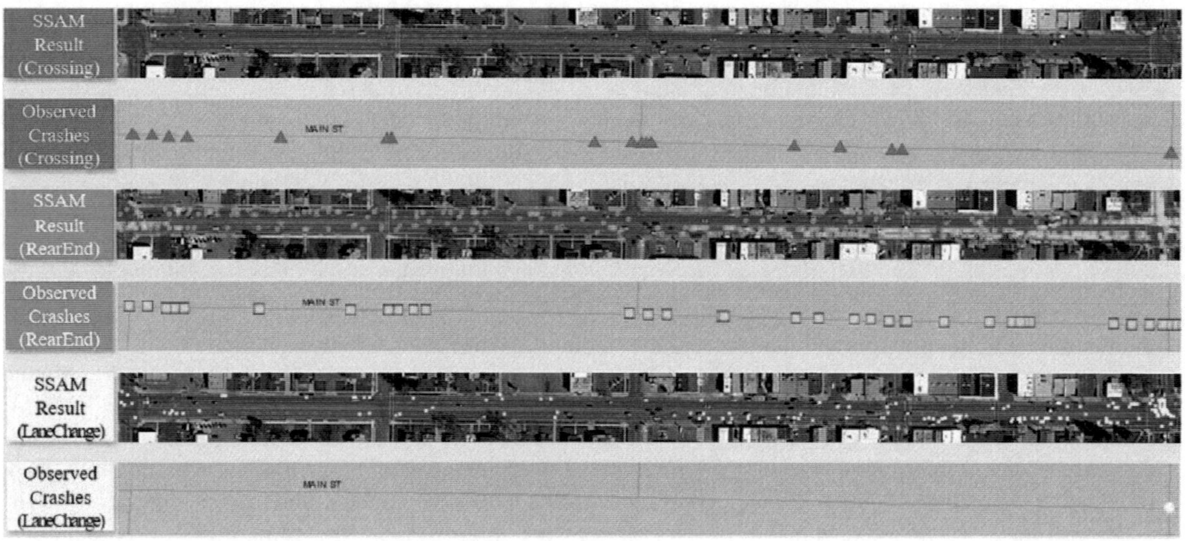

Figure 3. Spatial analysis of Segment 1 of the Main St. study section.

Calibration of SSAM for the Study Section

The simulation model was calibrated with 10 simulation runs before a comparative work was done. Table 3 shows the number of crashes from UDOT's crash data (*18*) and the number of conflicts from SSAM. Both crashes and conflict types were grouped into three groups: crossing, rear-end, and lane-change crashes or conflicts. To compare the trend in actual crash occurrences and conflict points determined by SSAM, a relation index was set up, which is the ratio between the number of actual crashes divided by the number of conflicts. This normalization was needed because a direct comparison of the numbers of crashes and the number of potential conflict points could not be made because of the difference in the timeframes of these values. Table 4 presents the relation indices calculated for the two segments. Overall, the variation in relation indices was small, indicating the stability of the SSAM-based analysis.

Figure 3 shows the spatial distributions of conflict points and actual crashes for Segment 1 of the study section. They are presented in pairs of three different conflict and crash types, namely crossing, rear-end, and lane-change crashes and conflicts. As shown in the figure, general trends in the location and concentration of crashes and conflicts are similar between the conflict points determined by SSAM and the crashes obtained from the crash data. Similar trends were found for Segment 2 (*3*).

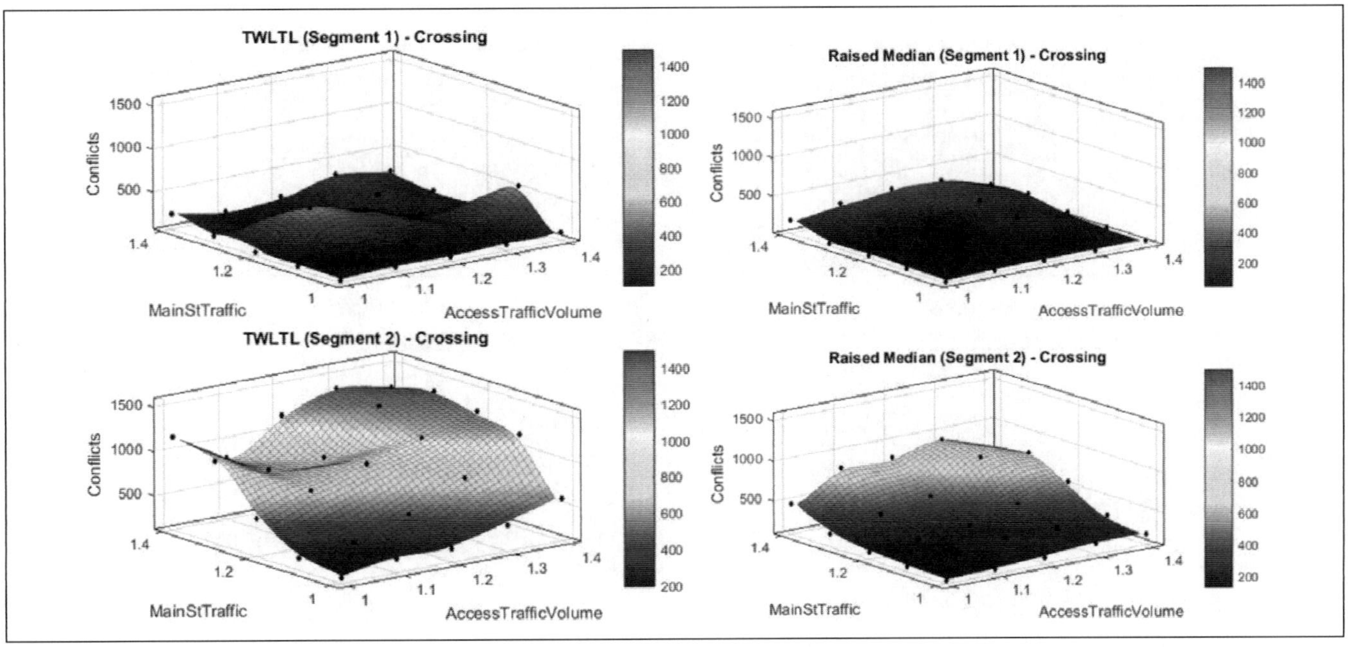

Figure 4. Three-dimensional presentation of changes in the number of crossing conflicts.

Use of SSAM for Evaluating Safety Impacts of Access Management Alternatives

With the SSAM calibrated, safety impacts of the two types of access management alternatives were compared. How much safety improvement can be made by installing a raised median? What would be the threshold of AADT to consider a raised median over a TWLTL? These are two typical questions traffic engineers would have in mind when considering replacing a TWLTL with a raised median. This exercise is an example of using SSAM as a tool to evaluate safety impacts of an access management alternative that can be implemented in the future.

Figure 4 shows a three-dimensional presentation of the trend in the changes in average number of crossing conflicts at different volume level combinations. To allow easier interpretation of the analysis results, a top-view projection of Figure 4 was created, which presents the number of conflict points in "contours" of crash frequency as shown in Figure 5. The top-view projection provides an opportunity to evaluate at what level of volume combinations the number of conflicts would change significantly. The top-view projection provided a clue for answering the second question posed in the previous paragraph. See Saito et al. (*3*) for the results for the other conflict types and total number of conflicts.

Since severe crashes tend to occur in crossing conflicts, the crossing conflict type is used to discuss general trends manifested by the conflict level contours. As shown in Figure 5, the numbers of crossing conflicts are quite similar between the TWLTL and the raised median in Segment 1, while they are significantly different between the two median types in Segment 2. The 2014 AADT of Segment 1 is 23,780 veh/day and AADT of Segment 2 is 29,870 veh/day, as shown in Table 1. Therefore, it can be said that at approximately 24,000 veh/day there is not much difference in safety-related benefits between these two alternatives even after volume increases. As AADT increases, however, the raised median begins to provide much safer traffic conditions compared with the TWLTL, as demonstrated in Segment 2 of the study site as shown in Figure 5. For example, the number of crossing conflicts for the TWLTL through Segment 2, with a 40 percent volume increase on Main St., ranges from 800 to 1,400, whereas the number of crossing conflicts for the raised median, with the same 40 percent volume increase on Main St., ranges from 400 to 900, a 32 percent to 50 percent reduction in crossing conflicts in Segment 2. Although the range of the number of conflict events for each conflict type differs among other conflict types, a similar trend was observed for rear-end and lane-change conflicts (*3*).

Results of the Sensitivity Analysis

Table 5 shows a brief summary of an analysis of variance (ANOVA) on crossing conflicts for Segment 1 and Segment 2. The ANOVA analysis showed that median type was the most dominant factor that would affect the number of conflicts followed by the volume on Main St., followed by the volume on access driveways. Although there were some interactions among the main factors, their contributions are much smaller than the three main factors. It was found that there was not much difference in the number of crossing conflicts in Segment 1, which had an AADT of 23,780 veh/day. In Segment 2, the differences in the number of crossing conflicts were significant, as shown in Figure 5. Segment 2 had

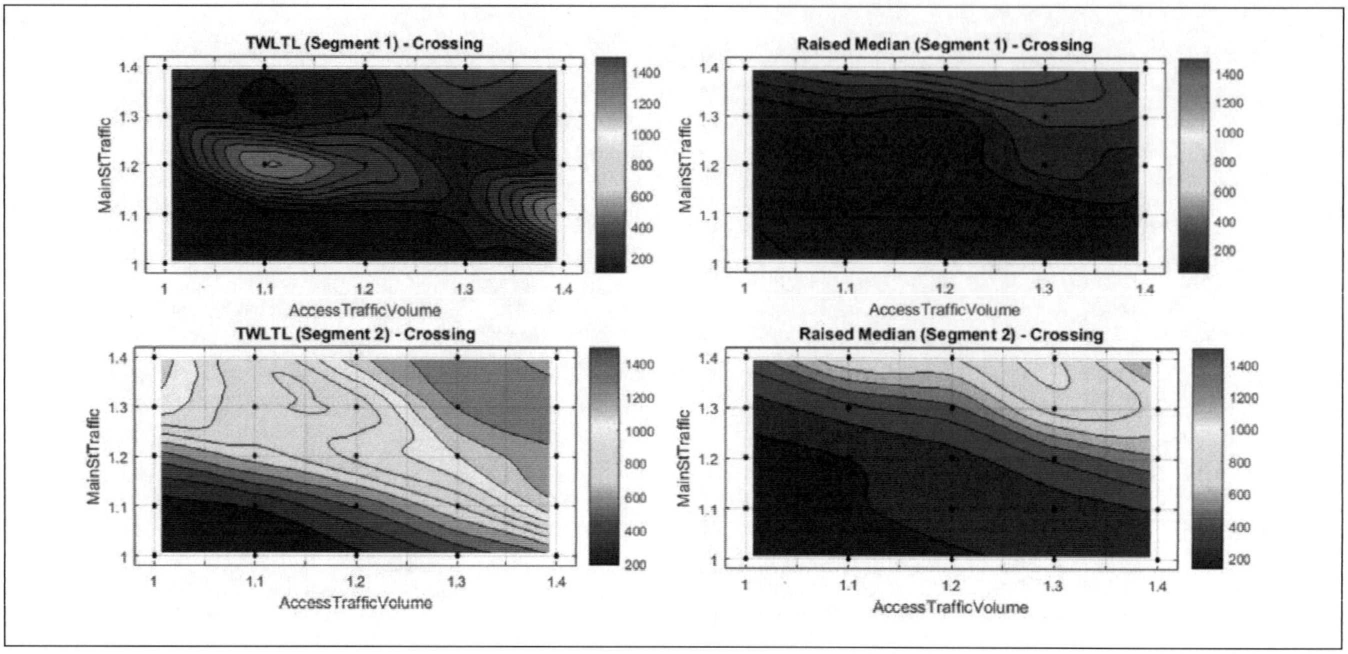

Figure 5. Top-view (contour maps) of crossing conflicts.

Table 5. ANOVA Test Results (Crossing Conflicts)

Effect	Number DF	Den DF	F-Value	Pr > F
Full model of Segment 1				
Median_Type	1	116	11.76	0.0008
Main Street Traffic Volume	4	116	3.92	0.0051
Access Driveway Traffic Volume	4	116	1.36	0.2538
Median_Type & Main Street Traffic Volume	4	116	1.83	0.1278
Median_Type & Access Driveway Traffic Volume	4	116	0.47	0.7572
Main Street & Access Driveway Traffic Volume	16	116	0.93	0.5348
Reduced model of Segment 1				
Main Street Traffic Volume	1	140	11.75	0.0008
Access Driveway Traffic Volume	4	140	3.91	0.0048
Median_Type & Main Street Traffic Volume	4	140	1.35	0.2530
Full model of Segment 2				
Median_Type	1	116	209.12	<0.0001
Main Street Traffic Volume	4	116	67.90	<0.0001
Access Driveway Traffic Volume	4	116	25.94	<0.0001
Median_Type & Main Street Traffic Volume	4	116	6.30	0.0001
Median_Type & Access Driveway Traffic Volume	4	116	5.08	0.0008
Main Street & Access Driveway Traffic Volume	16	116	1.99	0.0190
Reduced model of Segment 2				
Main Street Traffic Volume	1	140	151.38	<0.0001
Access Driveway Traffic Volume	4	140	49.15	<0.0001
Median_Type & Main Street Traffic Volume	4	140	17.78	<0.0001

an AADT of 29,870 veh/day. Note that both segments have similar access densities, as shown in Table 1.

To determine the threshold traffic volume for replacing a TWLTL with a raised median, cyan-colored contours were assumed as the upper threshold for using a TWLTL and the lower threshold to justify a raised median. This cyan-colored contour was selected as a potential threshold value because the cyan-color contour does not appear in Segment 1 either

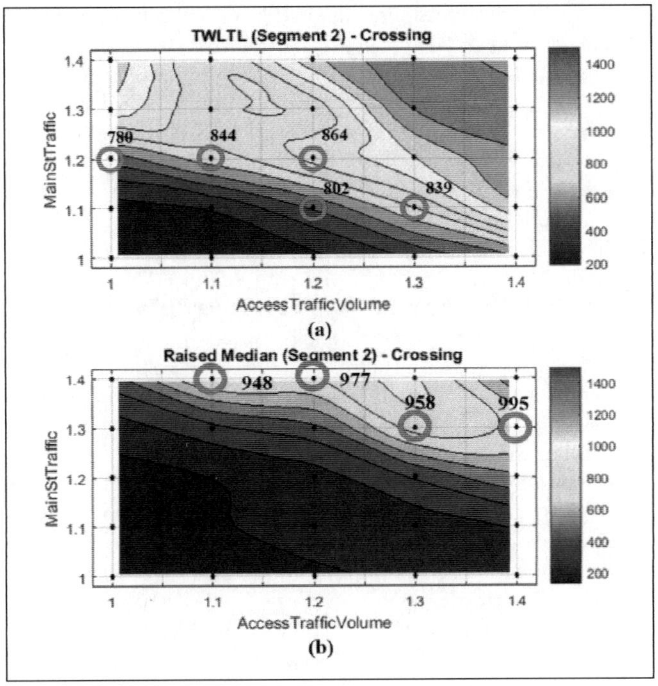

Figure 6. Threshold volume analysis results for crossing conflicts; (*a*) TWLTL median; (*b*) Raised median.

with a TWLTL median or with a raised median, as shown in Figure 5. This means that both access management measures have a similar performance in Segment 1 with AADT of approximately 24,000 veh/day. Because the most critical conflict type is crossing conflicts, it was decided to use the crossing conflict to find the threshold AADT to discuss the benefit of installing a raised median. In Figure 6, the cyan-colored contour represents the number of crossing conflicts of about 600. Therefore, the number of vehicles per lane for Main Street and its access demand combinations that are on or near the cyan-colored contour were determined.

Figure 6 shows the results of this threshold traffic-volume analysis for TWLTL and raised medians. The numbers shown in Figure 6 are the numbers of vehicles per hour per lane (veh/hr/ln) that would result in about 600 crossing conflicts for each segment. With the threshold of crossing conflicts in Segment 2 being set to approximately 600, traffic volumes on or near the cyan-colored contour were determined and they were found to be approximately 780 to 864 veh/hr/ln. With the raised median, traffic volumes on or near the cyan-colored contour were determined to be 948 to 995 veh/hr/ln. The raised median can accommodate 84 to 215 veh/hr/ln more to reach the same level of crossing conflicts for the TWLTL.

Knowing that the study segments have four lanes (a total of the number of lanes in both directions), the threshold traffic volume to convert a TWLTL to a raised median would be approximately 3,120 to 3,460 veh/hr ($780 \times 4 = 3,120$ and $864 \times 4 = 3,456$), as shown in Figure 6. Since the percentage of the current PM peak hour volume against AADT for the Main St. in the study area is about 9 percent (estimated using traffic volume in the VISSIM model provided by UDOT and AADT available from UDOT for year 2014 (*18*)), the threshold AADT for considering replacing a TWLTL with a raised median would be approximately 34,700 to 38,400 veh/day for the four lanes in Segment 2 (two lanes in each direction). A raised median may begin to have a similar level of crossing conflicts as the one for the TWLTL when the hourly volume becomes approximately 3,790 to 3,980 veh/hr ($948 \times 4 = 3,792$ and $995 \times 4 = 3,980$). With these hourly values, the AADT where a raised median might begin to have an increase in the number of crossing conflicts similar to the case with a TWLTL would be approximately 42,100 to 44,200 veh/day for Segment 2 of this study section.

As for safety improvement after replacing the TWLTL with the raised median, an approximate 32 to 50 percent reduction in the number of crossing conflicts was achieved when a raised median was used in lieu of a TWLTL when traffic volumes were increased by 40 percent of the base model at the Main St. study section. The percent reduction in crossing conflicts caused by changing a TWLTL with a raised median resembles the result of the Schultz et al. (*2*) study that was conducted in parallel to this study. The study compared crash occurrences before and after a change in median treatment from a TWLTL to a raised median at 20 study sites statewide, using Bayesian statistics. The study concluded that the median type change achieved crash reduction ranging from 32 to 44 percent for all severity groups except severity 4 and 5, for which a larger reduction of 57 percent and 58 percent, respectively, was achieved. These values are similar to the results of crossing conflict analysis.

Conclusion

With SSAM, a conflict analysis is done in a simulated environment; therefore, the physical and traffic characteristics of before and after can be maintained for before and after analyses. A planned access management alternative under study is modeled and no physical installation of the alternative is needed. Also, this conflict analysis is not affected by uncertainty inherent to crash data. The goal of this study was to evaluate if SSAM could be used to assess the safety impact of an access management alternative and to compare the safety impacts of multiple access management alternatives with less time, less cost, and less uncertainty.

In conclusion, the calibration and sensitivity studies conducted in this study showed that SSAM combined with a simulation model could be a viable tool to evaluate the safety impacts of access management alternatives planned for future implementation and a complement to traditional crash-based and in-field conflict analyses. In this study, conversion of a TWLTL into a raised median was evaluated to answer the two frequently asked questions related to such conversion: (1) How much safety improvement can be made by installing a raised median replacing a TWLTL? And (2) What would be the threshold of AADT to consider a raised median over a TWLTL?

The analysis showed that a raised median would be much safer than a TWLTL for the same level of traffic volume. An approximate 32 to 50 percent reduction in the number of crossing conflicts could be achieved. This is the answer to the first question mentioned in the paragraph above. Assuming a certain level of conflict occurrence level as a potential threshold to convert from a TWLTL to a raised median and the PM peak hourly volume percentage of 9 percent of AADT, average threshold daily traffic volumes in the study section were estimated. In this study, the threshold level was set to 600 crossing conflicts based on a typical trend observed in the outcomes of the analysis. The analysis showed that AADT of about 34,000 to 38,000 veh/day would be the demand level where a TWLTL is recommended to be converted into a raised median for the four-lane segments in the study segments. Also found was that the performance of a raised median might begin deteriorating once AADT becomes approximately 42,000 to 44,000 veh/day for the four-lane segments analyzed in this study. This is the answer to the second question mentioned in the paragraph above.

Acknowledgments

This research was made possible with funding from UDOT. Special thanks to Robert Miles, Tony Lau, Vincent Liu and the other members of the UDOT's Technical Advisory Committee for this study.

Author Contributions

The authors confirm contribution to the paper as follows: study conception and design: Kyung Min Kim, Mitsuru Saito, Grant G. Schultz; data collection: Kyung Min Kim, Mitsuru Saito; analysis and interpretation of results: Kyung Min Kim, Mitsuru Saito, Grant G. Schultz, Dennis L. Eggett; draft manuscript preparation: Kung Min Kim, Mitsuru Saito, Grant Schultz. All authors reviewed the results and approved the final version of the manuscript.

References

1. Williams, K., V. G. Stover, K. K. Dixon, and P. Demosthenes. *Access Management Manual*, 2nd ed. Transportation Research Board, Washington, D.C., 2014.
2. Schultz, G. G., M. L. Seat, and M. Saito. *Utilizing Lidar Data To Analyze Access Management Criteria In Utah*. Report No. UT-17.09. Research Division, Utah Department of Transportation, Salt Lake City, 2017.
3. Saito, M., K. Kim, and G. G. Schultz. *Analysis of Safety Impacts of Access Management Alternatives Using the Surrogate Safety Assessment Model*. Report No. UT-17.11. Research Division, Utah Department of Transportation, Salt Lake City, 2017.
4. Glennon, J. C., and B. A. Thorson. *Evaluation of the Traffic Conflicts Technique*. Office of Research and Development, Federal Highway Administration, U.S. Department of Transportation, Washington, D.C., 1975.
5. Gettman, D., and L. Head. *Surrogate Measures of Safety from Traffic Simulation Models*. Report No. FHWA-RD-03-050. Federal Highway Administration (FHWA), Washington, D.C., 2003.
6. Gettman, D., L. Pu, T. Sayed, and S. Shelby. *Surrogate Safety Assessment Model and Validation*. Final Report. Business Unit Intelligent Transportation Systems, Tucson, Ariz., 2008.
7. Gettman, D., and L. Head. *Surrogate Safety Measures from Traffic Simulation Models*. Final Report. Siemens Gardner Transportation Systems, Tucson, Ariz., 2012.
8. Sabra, Z. A., D. Gettman, D. Henry, and V. Nallamothu. *Balancing Safety and Capacity in an Adaptive Signal Control System – Phase 1*. Sabra, Wang & Associates, Inc, Baltimore, Md., 2010.
9. Hummer, F. E., R. L. Haley, S. E. Ott, R. S. Foyle, and C. M. Cunningham. *Superstreet Benefits and Capacities*. Department of Civil, Construction and Environmental Engineering, North Carolina State University, Raleigh, 2010.
10. Al-Ghandour, M. N., B. J. Schroeder, B. M. Williams, and W. J. Rasdorf. Conflict Models for Single-Lane Roundabout Slip Lanes from Microsimulation. *Transportation Research Record: Journal of the Transportation Research Board*, 2011. 2236: 92–101.
11. Lee, J., B. Park, K. Malakorn, and J. So. Safety Assessment of Cooperative Vehicle Infrastructure System-Based Urban Traffic Control. *Proc., 3rd International Conference on Road Safety and Simulation*, Indianapolis, IN. Transportation Research Board, 2011.
12. Huang, F., P. Liu, H. Yu, and W. Wang. Identifying if VISSIM Simulation Model and SSAM Provide Reasonable Estimates for Field Measured Traffic Conflicts at Signalized Intersections. *Accident Analysis and Prevention*, Vol. 50, 2013, pp. 1014–1024.
13. Dijkstra, A., P. Marchesini, F. Bijleveld, V. Kars, H. Drolenga, and M. van Maarseveen. Do Calculated Conflicts in a Micro-Simulation Model Predicting the Number of Crashes? *Transportation Research Record: Journal of the Transportation Research Board*, 2010. 2147: 105–112.
14. Pu, L., and R. Joshi. *Surrogate Safety Assessment Model (SSAM): Software User Manual*. Business Unit Intelligent Transportation Systems, Tucson, Ariz., 2008.
15. Stevanovic, A., J. Stevanovic, and C. Dergaye. Optimizing Signal Timings to Improve Safety of Signalized Arterials. *Proc., 3rd International Conference on Road Safety and Simulation*, Indianapolis, IN. Transportation Research Board, 2011.
16. Zhou, H., and F. Huang. Development of Traffic Safety Evaluation Method Based on Simulated Conflicts at Signalized Intersections. *Procedia–Social and Behavioral Sciences*, Vol. 96, 2013, pp. 881–885.
17. Habtemichael, F. G., and L. P. Santos. Crash Risk Evaluation of Aggressive Driving on Motorways: Microscopic Traffic Simulation Approach. *Transportation Research Part F: Traffic Psychology and Behaviour*, Vol. 23, 2014, pp. 101–112.
18. Utah Department of Transportation. *Open Data Portal*. http://udot.uplan.opendata.arcgis.com/. Accessed July, 2016.
19. Institute of Transportation Engineers. Trip Generation, 8th ed. Institute of Transportation Engineers, Washington, D.C., 2008.

The Standing Committee on Access Management (AHB70) peer-reviewed this paper (18-00280).

TRR
JOURNAL OF THE TRANSPORTATION RESEARCH BOARD

Article

Transportation Research Record
2018, Vol. 2672(17) 129–139

DOI: 10.1177/0361198118791864
journals.sagepub.com/home/trr
SAGE

Quantifying Access Management on Arterial Roadways

Lacy S. Brown[1,2], Karen K. Dixon[3], and H. Gene Hawkins, Jr.[1]

Abstract

Research suggests that successfully implemented access management programs can reduce delay, increase capacity, and improve safety performance on single roadway segments and across larger roadway networks. However, quantifying how access management, as a single entity, might affect a transportation system is difficult because countless combinations of strategies can be implemented. Consequently, large-scale access management decisions are often based on subjective assessments and the engineering judgment of practitioners and decision makers. There is a need for a standard, objective, and quantifiable approach to evaluating the impacts and performance of large-scale access management applications. This paper presents a quantitative method for evaluating an access management project on factors including operations; safety; impacts to adjacent land uses; and bicycle, pedestrian, and transit facilities. The result is an access management rating (AMR), a numerical value that allows for straightforward comparisons between corridors or between design alternatives on the same corridor. The proposed methodology eliminates the subjective component of the decision-making process while maintaining enough flexibility to be tailored to a specific agency's needs and priorities. By improving the consistency of access management evaluations, the decision-making process will be streamlined, funding will be allocated to projects with the greatest needs and opportunities for improvement, and the entire transportation industry will benefit from improved safety, operations, and land use development.

The literature indicates that implementation of access management strategies can reduce delay by up to 60%, increase capacity by up to 45%, and reduce crashes by up to 50% (*1*). Access management can be achieved through the application of any number of recommended strategies, many of which are well documented—in isolation. For example, previous research efforts have confirmed that a lower driveway density is associated with better safety and operational performance (*1*). Similarly, the presence of a raised median that restricts turning movements will likely result in fewer crashes and higher travel speeds than an undivided roadway (*2*, *3*).

But, imagine that you are looking to implement a combination of strategies on an arterial. Which strategies should you choose, and why? Would allowing a higher driveway density with a raised median perform better than a lower driveway density with no median? And, how is "better" defined? One might consider safety performance, travel speed, and bicycle and pedestrian usability as metrics, but what if the metrics provide inconsistent guidance? Which factors should be given more weight in the decision-making process?

These are the types of questions faced by access management professionals on a daily basis. Some practitioners and agencies develop criteria and metrics on a project-by-project basis, whereas others impart engineering judgment and subjective assessments, and others choose the easiest option or simply abandon the idea of access management altogether. This results in inconsistent applications of access management strategies and inefficient use of the limited available funding. There is a clear need for a standard, objective, and quantifiable approach to evaluating access management, especially for large-scale projects that incorporate multiple strategies and treatments.

This paper presents a quantitative method for evaluating an access management project based on a variety of factors including operations; safety; impacts to adjacent land uses; and bicycle, pedestrian, and transit facilities. This method can be customized to the priorities of a specific agency, providing a framework that is adaptable to

[1]Texas A&M University, College Station, TX
[2]DKS Associates, Salem, OR
[3]Texas A&M Transportation Institute, College Station, TX

Corresponding Author:
Address correspondence to Lacy S. Brown: lacy.brown@dksassociates.com

most access management project applications. The authors are confident that the overall access management assessment tool (AMAT) framework is applicable to a wide array of roadway types; however, the mathematical relationships derived for the AMAT are based on data from urban arterial corridors in the vicinity of highway interchanges.

Current Practice

A critical first step in developing a new methodology is to understand the current state of practice. The authors surveyed access management practitioners and general transportation engineers to gather insight into the decision-making processes currently applied when evaluating access management performance along a corridor. Additional information on the survey implementation and results can be found in the lead author's doctoral dissertation (*4*).

The survey consisted of a series of demographic questions followed by questions related to access management on six example corridors. Participants were shown (in random order) six arterial corridors including aerial and street-view photographs and site characteristics. An example of the survey information for Corridor E is presented in Figure 1.

The survey then presented each participant with the following questions.

1. How would you rate the level of access management of the given corridor on a scale of one to five (with five being excellent)?
2. Which of the following factors most influenced your rating selection?
    - Driveway density/spacing
    - Median type
    - Traffic signal density/spacing
    - Land use
    - Site development/circulation
    - Bicycle and pedestrian facilities
    - Transit facilities
    - Roadway geometry
    - Traffic volume
    - Other (please specify)

A total of 59 transportation professionals from the United States, Canada, Australia, and Poland completed the survey. The participants represented a variety of practice areas, employment types, and experience. Approximately 60% of participants either worked on access management projects regularly or considered themselves an expert in access management. These participants were assigned to the "high experience" category, and all others were included in the "low experience" category.

As expected, the results of the survey showed significant variation in the ratings assigned to each corridor and the factors identified as influencing the ratings. Even responses from those with extensive access management knowledge varied greatly (Figure 2).

Nearly the same number of "high experience" respondents rated the level of access management on Corridor E (Figure 1) with a score of one and a score of four, suggesting that respondents had vastly different opinions of what constitutes "good" access management. The ratings for the other five example corridors showed similar variability.

The survey results highlight the high level of subjectivity and variability involved in evaluating access management on a large scale. These types of evaluations will continue to be inconsistent unless a quantitative methodology is developed. Providing a means to objectively evaluate access management applications will enable practitioners to greatly improve the consistency with which corridors are planned, designed, retrofitted, and managed.

Overview of Methodology

This section provides an overview of the purpose, components, and end product of the proposed AMAT. The tool is intended to serve as a framework for access management evaluations; the individual components can be tailored to match an agency's priorities or modified to incorporate the findings of new research.

The primary goal of the AMAT is to provide a standard tool for conducting a quantitative assessment of access management that simultaneously considers multiple strategies and their combined impacts on the transportation system and surrounding land use.

The AMAT consists of three separate components that determine the operations, safety, and accessibility performance of a corridor based on the roadway and traffic characteristics linked to access management. The final access management rating (AMR) is determined by a weighted average of the three component ratings.

- **Operations Rating**—Mainline motor vehicle operations, quantified using corridor travel speed.
- **Safety Rating**—Corridor safety performance, quantified using crash risk.
- **Accessibility Rating**—The ease of accessing adjacent land uses by motor vehicles, bicyclists, pedestrians, and transit users, based on limited research and engineering judgment.
- **Resulting AMR**—Numerical score that represents the level of access management on a corridor, on a scale of 0.0–5.0.

Please review the following information and pictures describing the corridor, then answer the questions on the following page.

- Corridor Information: 5+ Lane Arterial
- Median: Center Two-Way Left-Turn Lane
- Posted Speed Limit: 40 mph
- Corridor Length: 0.43 miles
- Corridor AADT: 39,800 vpd

Aerial View of Corridor E

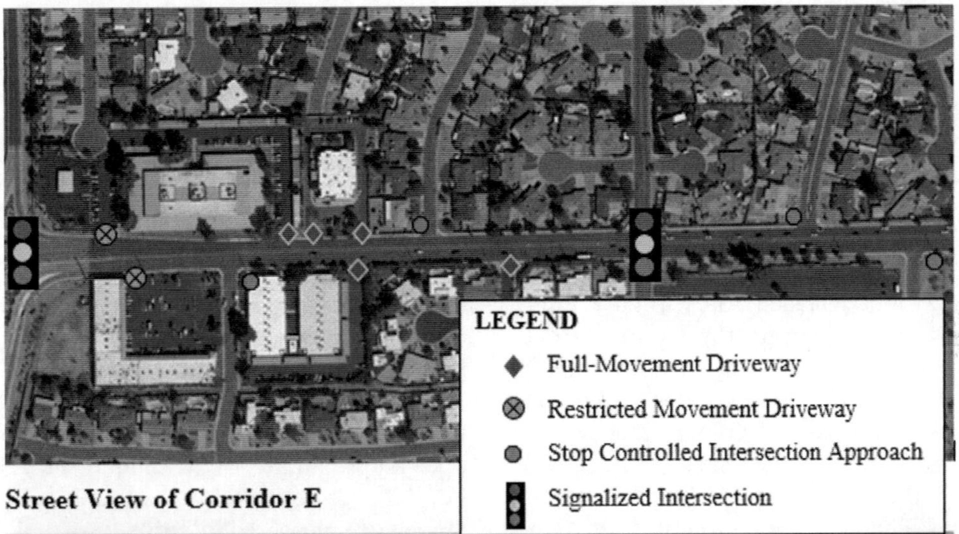

Street View of Corridor E

Figure 1. Survey information for Corridor E (Imagery © 2014 Google).

Development of the Methodology

The development of the AMAT included a combination of field-validated microsimulation modeling, statistical analysis, published research, and engineering judgment. Some components are data-driven whereas others rely on professional judgment because of a lack of available data and published research. The individual components were designed to allow for future refinement based on additional research. Although the authors are confident that the overall AMAT framework is applicable to a wide array of roadway types, the mathematical relationships derived for the AMAT most closely represent urban arterial corridors in the vicinity of highway interchanges because of the data used.

Field Data

The field data for the methodology came from study interchanges spanning four states (Arkansas, Arizona,

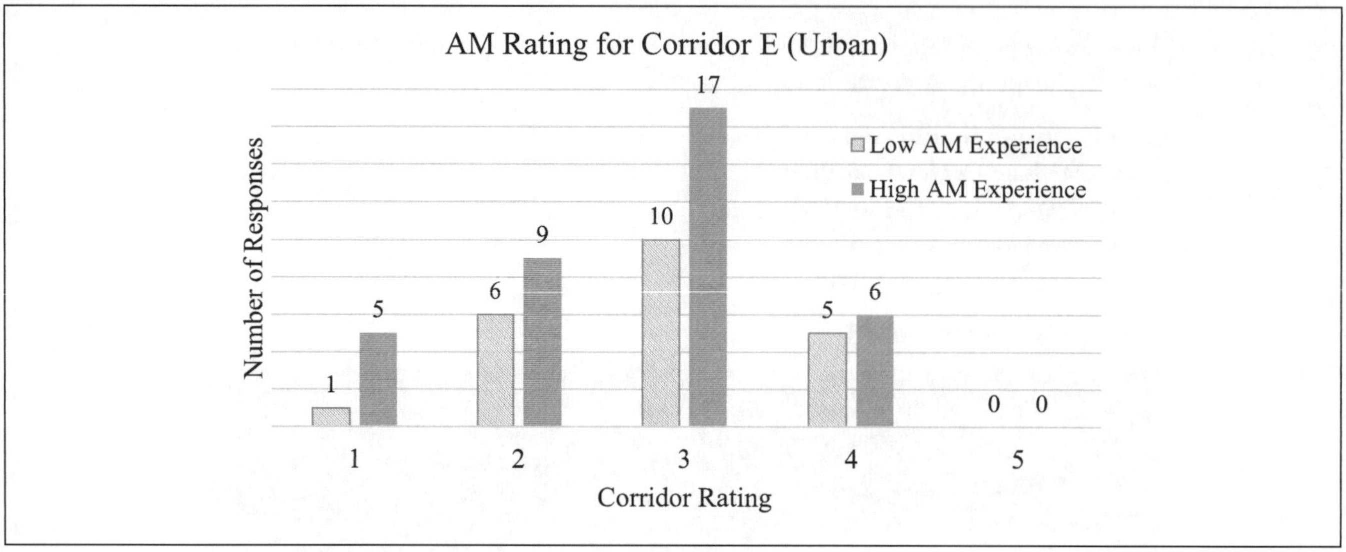

Figure 2. Access management ratings for Corridor E, by survey respondent experience.

Texas, and Virginia) and included four primary interchange forms (signalized diamond, stop controlled diamond, partial cloverleaf, and single point urban interchange). The 10 interchange sites yielded 20 potential study corridors (each side of an interchange was treated as an independent corridor). One interchange site was excluded from the operations analysis because of unique operational characteristics, resulting in nine usable sites and 18 study corridors.

The study corridors represented a wide variety of roadway characteristics, traffic volume, traffic control, land use, and level of access management. These corridors provided all of the field data used to develop the operations and safety components of the AMAT. Operations field data consisted of traffic volume, queue lengths, travel time and vehicle speeds, and signal timing information. Safety data consisted of 3 years of historical crash data, in the form of either individual crash reports or crash databases.

Operations Component

The foundation of the operations component of the AMAT was the development, validation, and testing of VISSIM microsimulation models. The authors modeled the existing site characteristics of each study corridor including roadway and intersection geometry, access location and geometry, traffic control, and traffic volume (mainline, intersection, and driveway volumes). Using corridor travel time as the determining metric, the authors successfully validated all 18 base models.

The authors then investigated the operational impacts of different access management strategies by developing alternative scenarios of each base model. The alternative models varied by median type, access point density, signalized intersection spacing, and the spacing between signalized intersections and adjacent driveways. Each combination of physical characteristics was also tested with varying levels of traffic. In total, the authors developed approximately 770 unique VISSIM models. The key data variables and value ranges associated with those models are presented in Table 1.

To quantify how different access management strategies affect operations (measured by corridor speed, averaged in both directions), the authors utilized regression analysis. Because the final urban data set consisted of 770 data points from only nine unique study locations, the data could not be analyzed under the assumption of independence. To account for this, the authors applied a mixed-effects modeling approach using a categorical variable identifying the study site as a random effect. The basis for model selection was a combination of Akaike information criterion (AIC), Bayesian information criterion (BIC), analysis of variance (ANOVA) comparisons, and graphical investigations of the data. The authors investigated the significance of the following explanatory variables.

- Posted Speed
- Median Type
- Interchange Type
- Distance between Signals
- Corridor Length
- Access Density
- Roadway Volume (vehicles per hour, or vph)
- Number of Lanes
- Volume per Lane (vehicles per hour per lane, or vphpl)

Table 1. Operations Data Variables and Value Ranges across Model Alternatives

Variable	Value range		
	Minimum value	Median value	Maximum value
Distance between signals (ft)	400	1,320	5,999
Corridor length (ft)	450	1,450	2,700
Signal cycle length (s)	95	130	180
Access density (per mile)	19	30	102
Mainline hourly volume (vph)	700	2,300	5,500
Daily volume (vpd)	12,150	40,500	104,400
Hourly volume per lane (vphpl)	133	467	925
Number of lanes	4	4	6
Posted speed (mph)	35	40	50
Average operating speed (mph)	6	23	37
Median type	Raised (continuous), raised (with strategic openings), two-way left-turn lane (TWLTL)		
Interchange functional form	Diamond, partial cloverleaf, single point urban		

Note: vph = vehicles per hour; vpd = vehicles per day; vphpl = vehicles per hour per lane.

- Validated Site Indicator
- Interchange Corner Clearance
- Percent Full Movement Driveways
- Percent Commercial Land Use
- Percent Residential Land Use
- Percent Vacant Land Use
- Percent Developed Land Use
- Driveways per Parcel
- Interactions of:
  - Posted Speed and Median Type
  - Median Type and Volume
  - Access Density and Median Type

Of the 20 explanatory variables investigated, three (median type, roadway volume, and distance between signalized intersections) were significant based on the regression analysis results. Surprisingly, neither the posted speed limit nor access density were significant predictors of corridor speed when median type was accounted for. Posted speed was a significant variable in models that excluded median type, but those models did not fit the data as well. This suggests that posted speed, median type, and access density are correlated to some degree, and median type is able to partially explain some of the impacts related to the other factors. The authors hypothesize that access density was not a significant variable because signal operations along the corridor had such a large impact on average speed that access characteristics were comparatively insignificant. Additional details on the microsimulation modeling and regression analysis can be found in the lead author's doctoral dissertation (*4*).

Although the microsimulation and regression results did not identify access density or spacing as a significant factor in predicting average corridor speed, there are multiple studies in the literature showing such a relationship does exist (*1*, *5*, *6*). These studies found very similar trends which indicate that every 10 additional access points per mile results in a corridor speed reduction of approximately 2.5 mph. The authors used this relationship to adjust the regression results such that the final Operations Rating (OR) equation also accounts for access density.

The authors wanted to account for the relative strength of the data used to develop the equation because they had full control of (and thus have high confidence in) the microsimulation and regression analysis, but had no involvement in the research discussed in the literature. To account for this, the authors used a weighted average of the predicted speeds calculated with and without the access density adjustment (weighting values of 0.6 and 0.4, respectively). The resulting equation for calculating the OR is

$$OR = 5 - \frac{8.31 - \left(0.002\text{Dsignal} - \text{MedTyp} - 0.015\text{vphpl} - \frac{\text{AccDen}}{10}\right)}{6.73} \quad (1)$$

where
OR = operations rating,
MedTyp = variable for median type (raised = 0, two-way left-turn lane [TWLTL] = 2.77, raised with strategic openings = 4.14),
Dsignal = distance between signalized intersections in feet (minimum 400, maximum 5,280),
vphpl = volume per hour per lane (minimum 150, maximum 1,000), and
AccDen = access points per mile (maximum 70).

Safety Component

The authors intended to conduct all safety analyses using the Federal Highway Administration's Surrogate Safety Assessment Model (SSAM), which can predict safety performance using a VISSIM model as input and would have allowed for the safety and operational analyses to be conducted on the exact same dataset. However, attempts to validate the SSAM results against actual crash data for three sites revealed that the SSAM software was not accurately estimating the safety performance of the study corridors. The authors revised the safety analysis approach to focus on the historical crash data for each study site, resulting in 20 data points with 3 years of data.

Using regression analyses to evaluate the safety performance of the study corridors requires the use of a mixed-effects model to account for the relation between each side of an interchange. Using this type of model would require a minimum of 100 data points (approximately 30 data points for each degree of freedom), significantly more data than was available for this study. The authors were therefore constrained to rely on descriptive statistics and trend analysis to evaluate the safety performance data of the study sites. This analysis, supplemented with the substantial body of literature on the topic, provides a solid foundation for the safety component of the AMAT.

The study sites, spanning four states, each have their own reporting threshold for property damage only (PDO) crashes, making total crash comparisons difficult. Therefore, the authors focused on fatal and injury (FI) crashes. The data indicated that crash rates were highest on roadway segments with a TWLTL. Segments with a raised median with strategic openings had higher crash rates than those with a continuous raised median. Crash rates generally increased as access density increased.

Because rigorous statistical analyses were not feasible for the relatively small data set of crash information, the development of the safety component of the AMAT relied more heavily on previously established relationships from the literature. The following relevant findings relate safety performance to access density and median type.

- Every additional driveway above 10 driveways per mile increases crash risk by approximately 4%. (*1, 7, 8*)
- Crash risk on roadways with TWLTLs is between 10% and 35% less than on undivided roadways. (*3, 7*)
- Crash risk on roadways with raised medians is between 15% and 57% less than on roadways with TWLTLs. (*1, 3, 9*)
- Crash risk on roadways with raised medians is up to 55% less than on undivided roadways. (*3*)
- Note: For purposes of AMAT development, the authors assume that the term "raised median" in the literature refers to a raised median with strategic openings (openings only at key driveways).

Two previous research efforts identified land use as a strong predictor of segment-intersection and total segment crashes. The authors used the regression results included in the more recent publication by Dixon et al. (*10*) to extrapolate a relationship between land use and safety performance. On urban roadway segments, the number of crashes is estimated to increase by approximately 17% for every 10% increase in the proportion of commercial and industrial driveways, after accounting for traffic volume, segment length, speed limit, median type, and number of lanes.

The authors used estimated crash rates as the measure of safety performance for developing the Safety Rating (SR) calculation equation. Using an arbitrarily selected base condition of 10 crashes per 100 million vehicle miles traveled, the authors developed an equation to account for the relationships found in the above safety analysis and the literature. Each additional access point per mile (above 10) increases the crash rate by 4%. Relative to a raised median with strategic openings, the crash rates are 20% lower for a continuous raised median, 30% higher for a TWLTL median, and 50% higher for no median. Finally, the crash rate increases by 17% for every 10% increase in the proportion of commercial and industrial driveways.

Multiple studies identified similar relationships between safety performance, median type, and access density, whereas only one study quantified the impact of land use on safety performance. The final SR calculation considers all of these relationships, and the difference in how extensively each of the relationships has been researched is accounted for by using a weighted average of the access and land use based relationships (weighting factors of 0.6 and 0.4, respectively). The resulting equation for calculating the SR is

$$SR = \frac{37.32 - 6 \times MedTyp \times (1 + .04(AccDen - 10)) - 4\left(1 + 1.7 \times \frac{\#ComIndDW}{Total\#DW}\right)}{5.7} \quad (2)$$

Table 2. Lookup Table for Accessibility Rating 1 (AR_1): Traffic Volume, Median Type, and Land Use

Median type	Proportion of developments with access to a minor roadway	Hourly traffic volume per lane (vphpl)			
		<600	601–800	800–1000	>1000
None	≤50%	2	1	0	0
	>50%	3	2	1	0
TWLTL	≤50%	3	2	1	0
	>50%	3	3	2	1
Raised with strategic openings	≤50%	4	3	3	2
	>50%	4	4	4	3
Continuous raised	≤50%	4	4	4	4
	>50%	5	5	5	5

where

SR = safety rating,

MedTyp = variable for median type (raised = 0.8, TWLTL = 1.3, raised with strategic openings = 1.0),

AccDen = access points per mile (minimum 10, maximum 70),

#ComIndDW = number of commercial and industrial driveways, and

Total#DW = total number of driveways.

Accessibility Component

Recognizing that the accessibility of transportation infrastructure to more modes of travel is receiving greater emphasis, the authors included a means of accounting for accessibility in the AMAT procedure. Because very little published research and no data currently exist for access management and accessibility, this component of the AMAT is primarily based on general guidelines in the literature and the engineering judgment of the authors. For purposes of the AMAT, the term "accessibility" is defined as the ease of access of land uses by all modes of travel (motor vehicles, bicycles, pedestrian, and transit users). Although the data may not be available to accurately quantify the relationship between access management and accessibility, the authors believe it is an important consideration that should not be ignored completely simply because sufficient data does not exist to quantify the relationships. The accessibility component of the AMAT includes the factors that authors judge are important and relevant. Additional research is required to confirm and quantify the relationships described below.

The accessibility rating (AR) incorporates two categories of accessibility, represented by AR_1 and AR_2. The variable AR_1 accounts for the potential impacts of access management on adjacent land use; AR_2 accounts for the presence of facilities for bicyclists, pedestrians, and transit users.

The variable AR_1 considers the impacts of a secondary street network and median type (a surrogate for driveway restrictions) on adjacent developments. The survey of practitioners suggested that the presence or lack of a secondary street network (which provides side-street access to developments) and interparcel connections are common considerations in evaluating access management along a corridor.

The economics-based literature indicates that access management projects, including median installations and driveway restrictions, have little to no negative impact on the property value or vitality of adjacent businesses (*11*, *12*). Additionally, a previous study of driveway restriction impacts on corner lot developments suggests that when a minor roadway access is present, providing unrestricted access to the major roadway does not reduce the amount of travel time required to enter and exit the development, particularly at higher roadway volumes (*13*). However, it is reasonable to assume that restricting the sole access point to a development will have a tangible impact on the travel patterns of development owners and patrons, even if that impact does not result in decreased business vitality. Because no quantitative data is available to estimate the precise accessibility impacts, the authors subjectively assigned the ratings using engineering judgment. The complete lookup table for AR_1 is presented in Table 2.

As shown in Table 2, a more extensive secondary network is associated with a higher rating because driveway restrictions and congestion levels will have a lesser impact on patrons needing to access adjacent developments because of the presence of alternative routes. Previous research has suggested that, depending on volume levels, unrestricted access is not necessarily beneficial to adjacent businesses even though business owners may claim otherwise (*13*). The authors selected the ratings in Table 2 to represent the trends more commonly associated with good access management practices (e.g., divided roadways are preferable to undivided roadways).

The second accessibility rating (AR_2) is the sum of three factors that account for pedestrian facilities (f_p), bicycle facilities (f_b), and transit facilities (f_t). A roadway

Table 3. Lookup Table for Pedestrian Facility Factor (f_p)

Percentage of corridor with continuous sidewalks	Signalized crossings per mile	Access density (access points per mile)			
		<10	10–24	25–40	>40
>75%	≥4	3	2	1	1
	<4	2	1	1	0.5
50–75%	≥4	2	1	0.5	0.5
	<4	1	0.5	0.5	0
<50%	≥4	1	0.5	0	0
	<4	0.5	0	0	0

Table 4. Lookup Table for Bicycle Facility Factor (f_b)

Bicycle facilities present	Roadway speed	
	≤35 mph	>35 mph
None	0	0
Wide paved shoulder	0.5	0.5
Shared lane (marked)	0.5	0
On-street bicycle lane	1.0	1.0
Off-street bicycle or shared use path	1.0	1.0

with the "best case" conditions for all three factors has a combined accessibility rating (AR_2) of five. Because there are significantly more pedestrians than bicyclists or transit riders, particularly in the U.S., the authors assigned a maximum value of three to f_p, and a maximum value of one to both f_b and f_t. As stated previously, the ratings for f_p, f_b, f_t, and AR_2 are primarily based on the authors' engineering judgment because of a lack of available data. The process for calculating AR_2 is summarized as

$$AR_2 = f_p + f_b + f_t \quad (3)$$

where

AR_2 = accessibility rating 2,
f_p = pedestrian facility factor,
f_b = bicycle facility factor, and
f_t = transit facility factor.

The pedestrian facilities factor, f_p, takes into account the presence of continuous sidewalks, the density of signalized crossings, and access density (Table 3). The *Access Management Manual* (*1*) recommends limiting the number of driveways to reduce the number of vehicle–pedestrian conflicts and providing continuous pedestrian facilities where pedestrian traffic is expected. Additionally, the *Highway Capacity Manual* (*6*) pedestrian level of service (LOS) procedure penalizes the pedestrian score when midblock crossings are increasingly difficult (higher number of lanes) and when significant out of direction travel is required (dedicated crossing locations are far apart). The authors reviewed pedestrian planning guidelines from Oregon, Washington, and Florida, but could not find any recommendations for minimum crosswalk spacing. To maintain consistency with the *Access Management Manual*'s minimum signal spacing recommendation of one-quarter mile, the authors used four signalized crosswalks per mile as the threshold between categories.

The bicycle facilities factor, f_b, takes into account the type of bicycle facilities present and the roadway speed. The *Access Management Manual* (*1*) recommends providing dedicated bicycle facilities in urban areas, but suggests that wide, high-quality paved shoulders can be used in rural areas. Additionally, the Manual on Uniform Traffic Control Devices (*14*) states that shared lanes (marked with a sharrow) should not be used on roadways with a speed limit above 35 mph. The bicycle facilities factor lookup table is presented in Table 4.

The final variable in the calculation of the variable AR_2 is the transit facilities factor, f_t. The *Transit Cooperative Research Program Report 19* (*15*) provides recommendations for the density of transit stops in urban and suburban areas that directly correlate to the values of f_t where $f_t = 0$ if no transit stops are present, $f_t = 1.0$ if there are more than five transit stops per mile, and $f_t = 0.5$ for intermediate densities.

The Access Management Rating (AMR)

AMR Calculation

The calculation of the AMR utilizes the above lookup tables and equations to yield a single value representing the level of access management. The AMR is a decimal value on a scale of zero to five, with zero representing poor (or no) access management and five representing excellent access management. Equation 4 presents the AMR calculation based on the operations, safety, and accessibility considerations described in this chapter.

$$AMR = A_1(OR) + A_2(SR) + A_3(AR_1) + A_4(AR_2) \quad (4)$$

where

AMR = access management rating,
A_1, A_2, A_3, A_4 = adjustment factors,
OR = operations rating,
SR = safety rating, and
AR_1, AR_2 = accessibility and land use ratings.

The adjustment factors shown in Equation 4 provide a means to calculate a weighted average of the operations, safety, and accessibility considerations included in the AMAT. The values shown in Table 5 represent the relative importance of operations, safety, land use accessibility, and nonmotorized accessibility, respectively,

Table 5. Lookup Table for Adjustment Factors (A_1, A_2, A_3, A_4)

Roadway type	Future development plan	A_1 (Ops.)	A_2 (Safety)	A_3 (Land Use)	A_4 (Bike, Ped., & Transit)
New construction (undeveloped)	Low–medium density land use	0.40	0.40	0.10	0.10
	Medium–high density land use	0.40	0.40	0.05	0.15
Existing (full or partial development)	Low–medium density land use	0.35	0.35	0.15	0.15
	Medium–high density land use	0.35	0.35	0.10	0.20

Note: Ops. = operations; Ped. = pedestrian.

for different levels of existing and future development. Table 5 presents the recommended values for adjustment factors A_1, A_2, A_3, and A_4, which should always sum to a value of 1.0. In all cases, operations and safety (A_1 and A_2) are equally weighted. Considerations for accessibility (A_3 and A_4) have higher weights for locations where existing development is present, as current land use owners and patrons will be directly affected by access management decisions. As the level of future planned development increases, the relative consideration of land use accessibility (A_3) decreases (to accommodate high levels of development, roadway safety and mobility must be a priority) whereas the relative consideration of bicycle, pedestrian, and transit accessibility (A_4) increases (to promote the safety and mobility of nonvehicular traffic). These are suggested starting values that should be adjusted to represent an agency's unique priorities for operations, safety, and accessibility.

Example Applications

The AMAT can be applied to any level of access management decision making, from treatment selections on small projects to agency-wide funding priorities. Two likely applications are in the network screening process and the selection of preferred design alternatives. For network screening, an agency may select an acceptable AMR value and flag those corridors within the network with lower AMR values for further study. Once studied, the relative AMR values can also be used to prioritize potential improvements. For alternative design selection, the AMR can provide useful information on the tradeoffs between safety, operations, and accessibility when looking at various design options for a new or reconstructed corridor.

Figure 3 presents an overview of the AMR calculation results for two urban arterial roadways with different levels of access management.

Summary

The literature review highlighted a lack of research documenting the combined impacts (on safety, operations, and accessibility) of implementing multiple access management strategies simultaneously. As a result, practitioners must currently impart subjective assessments and engineering judgment when making access management decisions. The AMAT procedure fills the need for a consistent, objective, and quantifiable means of evaluating access management impacts and performance on a large scale. Using readily available data to provide a project-level rating value, the AMR allows for straightforward comparisons between corridors or between design alternatives for the same corridor. The authors envision this tool being utilized by state and local agencies for network screening purposes by setting an AMR target value and flagging corridors that do not meet that threshold. Additionally, design engineers and access management professionals can use the AMAT to compare the access management performance of different design alternatives for a new corridor or the reconstruction of an existing corridor. The AMAT framework incorporates substantial flexibility to refine and customize the individual components without compromising the quantitative nature of the tool. Application of the AMAT procedure has the potential to improve the consistency and accuracy with which access management decisions are made within the transportation profession.

Limitations and Future Work

The AMAT is the result of the first known effort to provide a quantitative method for evaluating access management on a project level. The authors see this work as a strong foundation for standardizing the future of access management evaluations. The AMAT represents a significant advancement from the current state of practice, and the authors acknowledge several opportunities for improving and refining the AMAT procedure.

- The data used to develop the AMAT was limited to major arterials close to a grade-separated interchange. Conducting further research on roadway corridors with different characteristics and contexts would improve the validity of the tool.
- Expanding the data set to include rural considerations would broaden the applicability of the tool. Alternatively, a complementary AMAT procedure for rural roadways could be developed.
- The authors relied on a very limited body of research and a relatively large amount of

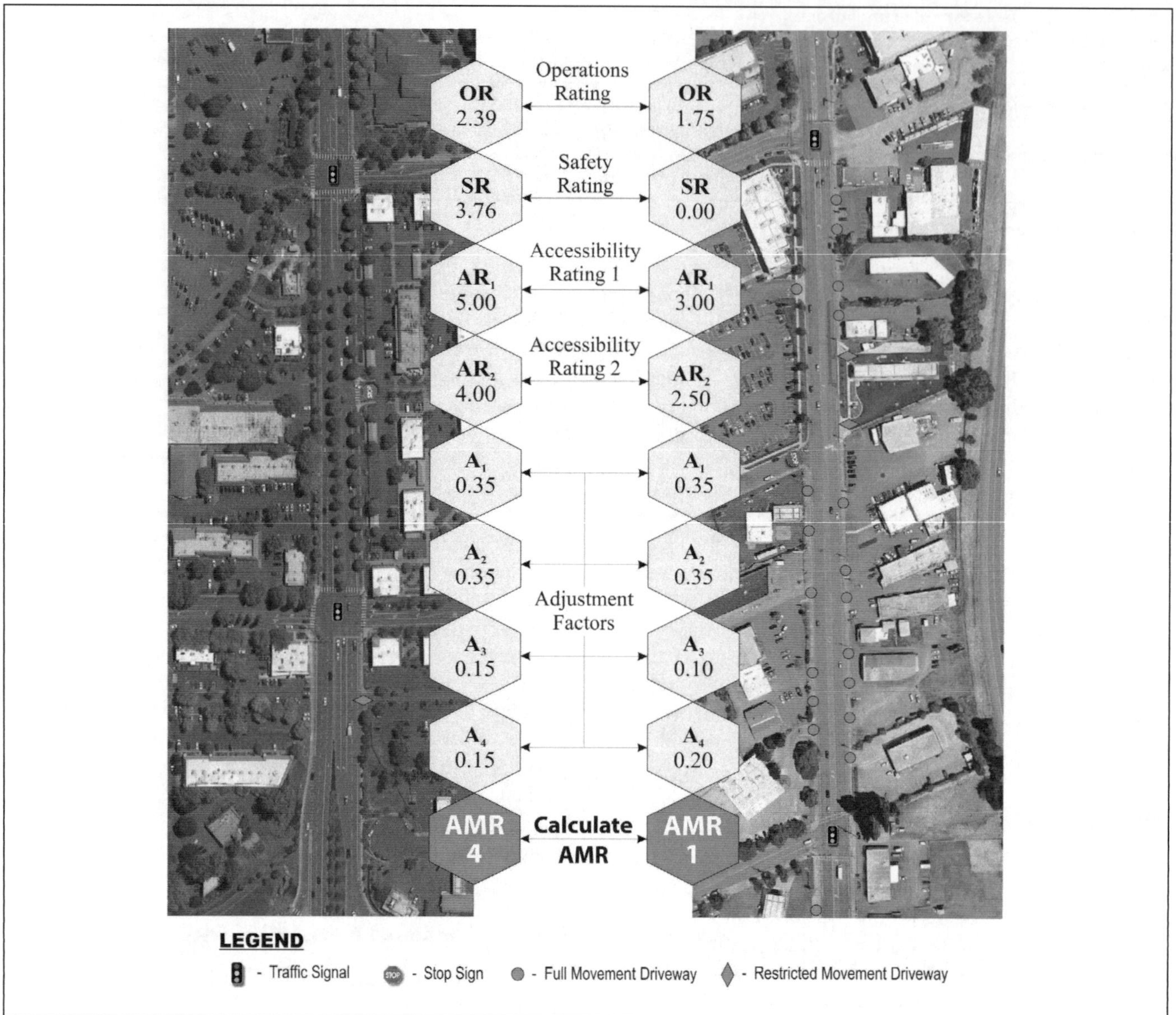

Figure 3. Example AMR calculations (Imagery © 2017 Google).

engineering judgment in estimating the land use, pedestrian, bicycle, and transit impacts of access management strategies. Any additional research on these topics would significantly improve the strength of the AMAT methodology.

Acknowledgments

This research was conducted as part of the lead author's doctoral dissertation. Funding for the dissertation included FHWA's Eisenhower Transportation Fellowship. The field data used to develop this methodology was obtained with permission from NCHRP and the project oversight panel as part of NCHRP Project 07-23 (Access Management in the Vicinity of Interchanges).

Author Contributions

The authors confirm contribution to the paper as follows: study conception and design: Lacy S. Brown, Karen K. Dixon, and H. Gene Hawkins, Jr.; data collection: Lacy S. Brown, Karen K. Dixon, and H. Gene Hawkins, Jr.; analysis and interpretation of results: Lacy S. Brown, Karen K. Dixon, and H. Gene Hawkins, Jr.; draft manuscript preparation: Lacy S. Brown, Karen K. Dixon, and H. Gene Hawkins, Jr.. All authors reviewed the results and approved the final version of the manuscript.

References

1. Williams, K., V. Stover, K. Dixon, and P. Demosthenes. *Access Management Manual*, 2nd ed. Transportation

Research Board of the National Academies, Washington, D.C., 2014.
2. Bonneson, J., and P. McCoy. *NCHRP Report 395: Capacity and Operational Effects of Midblock Left-Turn Lanes.* Transportation Research Board of the National Academies, Washington, D.C., 1997.
3. S&K Transportation Consultants. *Access Management, Location, and Design.* NHI Course 133078 Participant Notebook. Federal Highway Administration, Washington, D.C., 2000.
4. Brown, L. S. *A Method for Evaluating Access Management on Arterial Crossroads in the Vicinity of Interchanges.* Doctoral dissertation. Texas A&M University, College Station, 2015.
5. McShane, W., D. Choi, K. Eichin, and G. Sokolow. Insights into Access Management Details Using TRAF-NETSIM. *Proc., National Conference on Access Management*, Federal Highway Administration, Vail, Colo., 1996, pp. 471–483.
6. *Highway Capacity Manual 2010.* Transportation Research Board of the National Academies, Washington, D.C., 2010.
7. Gluck, J., H. Levinson, and V. Stover. *NCHRP Report 420: Impacts of Access Management.* Transportation Research Board of the National Academies, Washington, D.C., 1999.
8. Papayannoulis, V., J. S. Gluck, K. Feeney, and H. S. Levinson. Access Spacing and Traffic Safety. *Proc., Urban Street Symposium*, Dallas, Texas, 1999.
9. Parsonson, P., M. Waters, and S. Fincher. Effect on Safety of Replacing an Arterial Two-Way Left-Turn Lane with a Raised Median. *Proc., 1st National Access Management Conference*, Federal Highway Adminsitration, Vail, Colo., 1993, pp. 265–270.
10. Dixon, K., R. Avelar, L. Brown, M. Mecham, and I. van Schalkwyk. *Quantifying Safety Performance of Driveways on State Highways.* Oregon Department of Transportation, Salem, 2012.
11. Benz, R., N. Norboge, A. Voigt, and S. Gage. An Economic Assessment of Access Management Projects in the Houston Region. *Transportation Research Record: Journal of the Transportation Research Board*, 2015. 2486: 80–89.
12. Eisele, W., and W. Frawley. *A Methodology for Determining Economic Impacts of Raised Medians: Data Analysis on Additional Case Studies.* Texas Department of Transportation, Austin, 1999.
13. Brown, L., and K. Dixon. Use of Access Travel Time to Estimate the Impact of Driveway Restrictions on Corner Lot Developments. *Transportation Research Record: Journal of the Transportation Research Board*, 2015. 2486: 74–79.
14. *Manual on Uniform Traffic Control Devices.* Federal Highway Administration, Washington, D.C., 2009.
15. Texas Transportation Institute. *TCRP Report 19: Guidelines for the Location and Design of Bus Stops.* TRB, National Research Council, Washington, D.C., 1996.

The Standing Committee on Access Management (AHB70) peer-reviewed this paper (18-01612).